Chemical Reagents for Protein Modification

Volume II

Authors

Roger L. Lundblad, Ph.D.
Professor of Pathology
and Biochemistry
Dental Research Center
University of North Carolina
Chapel Hill, North Carolina

Claudia M. Noyes, Ph.D.
Research Associate
Department of Medicine
University of North Carolina
Chapel Hill, North Carolina

CRC Press, Inc.
Boca Raton, Florida

TP
453
. P7
L86
1984
vol. 2

Library of Congress Cataloging in Publication Data

Lundblad, Roger L.
    Chemical reagents for protein modification.

    Bibliography: p.
    Includes index.
    1. Proteins. 2. Chemical tests and reagents.
I. Noyes, Claudia M. II. Title.
TP453.P7L86   1984   661'.8   83-15076
ISBN 0-8493-5086-7 (v. 1)
ISBN 0-8493-5087-5 (v. 2)

© 1984 by CRC Press, Inc.
Second Printing, 1984
Third Printing, 1985
Fourth Printing, 1985
Fifth Printing, 1988
Sixth Printing, 1989

International Standard Book Number 0-8493-5086-7 (v. 1)
International Standard Book Number 0-8493-5087-5 (v. 2)

Library of Congress Card Number 83-15076
Printed in the United States

PREFACE

The contents of this book are focused on the use of chemical modification to study the properties of proteins in solution. Particular emphasis has been placed on the practical laboratory aspects of this approach to the study of the relationship between structure and function in the complex class of biological heteropolymers. As a result, little emphasis is given to the individual consideration of the functional consequences of chemical modification.

The authors are indebted to the many investigators who have allowed us to include their data in this book. Such cooperation has permitted us to discuss many aspects of the specific modification of protein molecules. We are also indebted to the many journals that have allowed us to reproduce copywritten material from many laboratories.

The authors wish to express their graditude to Ms. Reneé Williams for help in the preparation of the text and to Ramona Hutton-Howe, Terri Volz, David Rainey and their colleagues in the Learning Resources Center of the School of Dentistry at the University of North Carolina at Chapel Hill for preparation of the material for figures.

One of the authors (RLL) wishes to express his particular gratitude to past and present colleagues on the fifth floor of Flexner Hall at the Rockefeller University for their contributions to this book as well to the considerable support provided for other endeavors.

The authors' research programs are supported by grants DE-02668, HL-06350 and HL-29131 from the National Institutes of Health.

Roger L. Lundblad
Claudia M. Noyes

THE AUTHORS

Roger Lauren Lundblad, Ph.D. is Professor of Pathology and Biochemistry in the School of Medicine and Professor of Oral Biology in the Department of Periodontics in the School of Dentistry at the University of North Carolina at Chapel Hill. He also serves as Associate Director for Administration of the Dental Research Center.

Dr. Lundblad received his undergraduate education at Pacific Lutheran University in Tacoma, Washington and the Ph.D. degree in biochemistry at the University of Washington in 1965. Prior to joining the faculty of the University of North Carolina in 1968, Dr. Lundblad was a Research Associate at the Rockefeller University.

Dr. Lundblad's research interests are in the use of solution chemistry techniques to study protein-protein interaction with particular emphasis on the proteolytic enzymes involved in blood coagulation and the proteins synthesized and secreted by salivary glands.

Claudia Margaret Noyes, Ph.D. is a Research Associate in the Department of Medicine at the University of North Carolina at Chapel Hill.

Dr. Noyes received her undergraduate education at the University of Vermont and the Ph.D. degree in chemistry from the University of Colorado in 1966. Prior to joining the faculty at the University of North Carolina at Chapel Hill, Dr. Noyes was a Research Associate at the University of Chicago and a Research Chemist at Armour-Dial.

Dr. Noyes's current research interests include structural studies of proteins involved in blood coagulation and high performance liquid chromatography of peptides and proteins.

TABLE OF CONTENTS

Volume I

TABLE OF CONTENTS

Volume II

Chapter 1

THE MODIFICATION OF ARGININE

Until approximately 10 years ago the specific chemical modification of arginine was relatively difficult to achieve. The high pKa of the guanidine functional group (pKa $\simeq$ 12 to 13) necessitated fairly drastic reaction conditions (pH $\geq$ 12) to generate an effective nucleophile. Most proteins are not stable to extreme alkaline pH. The modification of arginyl residues was however possible and the early efforts in this area have been previously reviewed.[1]

It is reasonable to suggest that the recent advances in the study of the function of arginine residues in proteins stem from the work of Takahashi[2] on the use of phenylglyoxal as a reagent for the specific modification of arginine although observations on the use of 2,3-butanedione[3] and glyoxal[4] appeared at approximately the same time. The greatly increased interest in the elucidation of functional arginyl residues probably arises from the suggestion of Riordan and co-workers[5] that arginyl residues function as "general" anion recognition sites in proteins. Patthy and Thész[6] extended this concept by suggesting that the pKa of arginyl residues at anion binding sites (Figure 1) is lower than that of other arginine residues which would explain the specificity of the dicarbonyl residues which will be discussed below.

This chapter will primarily consider the reaction of arginyl residues in proteins with three different reagents; phenylglyoxal, 2,3-butanedione, and 1,2-cyclohexanedione since the vast majority of reports during the past decade have used these reagents. It is noted that several other reagents have been used for the modification of arginine. The modification of arginyl residues with ninhydrin occurs under relatively mild conditions (pH 8.0, 25°C, 0.1 M N-ethylmorpholine acetate, pH 8.0) as described by Takahashi.[7] The modification of pancreatic ribonuclease A or ribonuclease T by ninhydrin is shown in Figure 2. The reaction proceeds quite rapidly at pH 8.0 with the modification of both arginyl and lysyl residues. Reducing the pH to 5.5 (0.1 M sodium acetate, pH 5.5) reduced the rate of inactivation but did not increase the specificity of the modification. The UV spectra of ribonuclease T, before and after modification with ninhydrin are presented in Figure 3. Takahashi[7] achieved specificity of modification by first modifying available lysyl residues with a reagent such as methylmaleic anhydride (citraconic anhydride) which can subsequently be removed under conditions where the arginine derivative is stable (pH 3.6). The arginine derivative is unstable under basic conditions (1% piperidine, ambient temperature, 34 hr) and arginine was regenerated. Under the conditions commonly used for the preparation of protein samples for amino acid analysis (6 N HCl 110°C, 24 hr), the ninhydrin-arginine derivative was destroyed with the partial regeneration of free arginine. The structure of the ninhydrin-arginine derivative (Figure 4) is similar to that proposed for the α,α'-dicarbonyl compounds such as phenylglyoxal[2] or 1,2-cyclohexanedione.[8] At the same time a report by Chaplin[9] appeared proposing the use of ninhydrin for the reversible modification of arginine residues in proteins. The study suggested that at pH 9.1(0.1 M sodium phosphate), 37°C, the rate of reaction at arginine residues is approximately 100-fold more rapid than at lysine residues but reaction at cysteinyl residues is approximately 100-fold more rapid than at arginine. The extent of reaction was determined by measuring the decrease in the absorbance of ninhydrin at 232 nm (ϵ = 3.4 $\times$ 10^4 M^{-1} cm^{-1}). As noted by Takahashi,[7] the ninhydrin-arginine derivative is unstable under alkaline conditions and can be used for the reversible modification of arginine residues. The fluorescence properties of the reaction product between ninhydrin and guanidino compounds such as arginine have provided the basis for the use of ninhydrin for the detection of guanidine compounds in biological fluids (plasma) following separation by high performance liquid chromatography.[10]

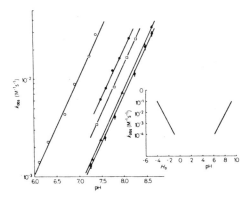

FIGURE 1.   Dependence of the reaction of arginine with dicarbonyl compounds on pH. Shown are second-order rate constants for the reaction of phenylglyoxal (○———○), hydroxypyruvaldehyde (□———□), glyoxal (▲———▲), and 1,2-cyclohexanedione (♦———♦) with free arginine at 25°C at the indicated pH (the buffers were 0.1 M sodium phosphate for pH 6 to 8; 0.1 M triethanolamine-HCl for pH 7 to 9 and in HCl solutions (H_o-4-0). Also included are the rate constants for the reaction of 1,2-cyclohexanedione with aldolase determined in 0.1 M triethanolamine-HCl buffers at 25°C (●———●). The inset shows the second-order rate constants for the arginine-glyoxal reaction over a wider pH range. (From Patthy, L. and Thész, J., *Eur. J. Biochem.*, 105, 387, 1980. With permission.)

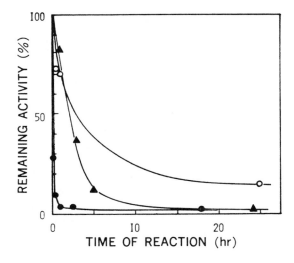

FIGURE 2.   Rates of inactivation of ribonucleases A and T_1 by ninhydrin. The reaction was performed at 25°C in the dark either at pH 8.0 in 0.1 M N-ethylmorpholine acetate, or at pH 5.5 in 0.1 M sodium acetate, at a protein concentration of 0.5% and a ninhydrin concentration of 1.5%. pH 8.0: ●, ribonuclease A; ▲, ribonuclease T_1. pH 5.5: ○, ribonuclease A. (From Takahashi, K., *J. Biochem.*, 80, 1173, 1976. With permission.)

The modification of arginyl residues with glyoxal has also been proposed.[11] Specificity of reaction is a problem with reaction also at primary amine groups and sulfhydryl groups. For example, reaction of glyoxal with bovine serum albumin at pH 9.0 resulted in modification of greater than 80% of the arginine residues with approximately 30% modification of lysine residues.[11] Glass and Pelzig[12] have examined the reversible modification of arginyl residues with glyoxal in some detail. Several products are formed from the reaction of

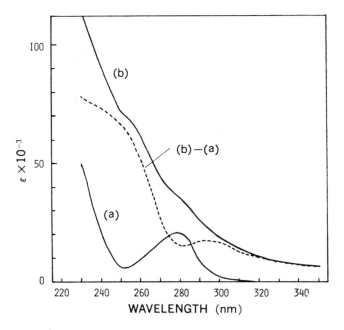

FIGURE 3. Changes in the UV absorption spectrum of ribonuclease T₁ on reaction with ninhydrin. The spectra were measured in 0.01 *M* ammonium acetate. ———: (a) native ezyme; (b) ninhydrin-modified enzyme. ----: difference, (b) − (a). (From Takahashi, K., *J. Biochem.*, 80, 1173, 1976. With permission.)

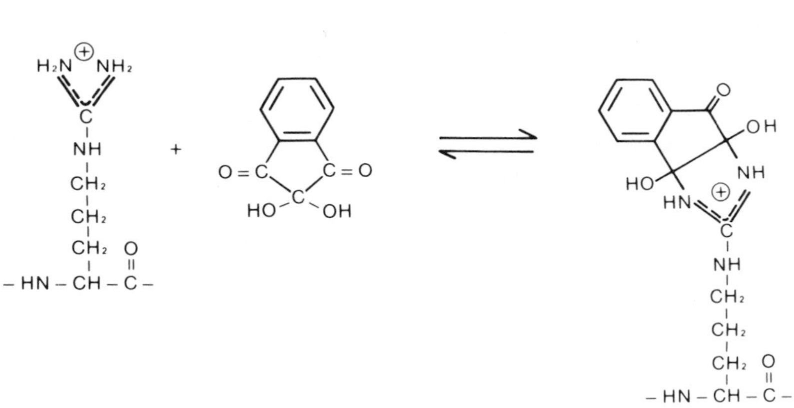

FIGURE 4. A scheme for the reaction of ninhydrin with arginine.

glyoxal and arginine at alkaline pH. One of these derivatives is markedly stable in strong acid(12 *M* HCl) at ambient temperature but rapidly is degraded to form free arginine in the presence of *O*-phenylenediamine (0.16 *M*) at pH 8.1 to 8.3. More alkaline conditions resulted in more rapid decomposition of the glyoxal-arginine derivative and ninhydrin-positive compounds other than arginine were formed. Reaction of arginine with glyoxal in borate buffer also yields the product described above. The same research group has reported on the reversible modification of arginine residues with camphorquinone-10-sulfonic acid and derivatives such as camphorquinone-10-sulfonylnorleucine.[13] The synthesis of the parent com-

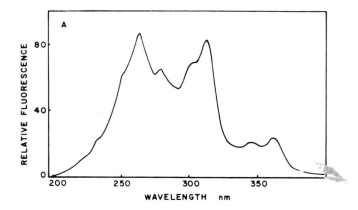

FIGURE 5.   The fluorescence excitation spectrum of the condensation product of arginine and 9,10-phenanthrenequinone. Arginine (90 μg/mℓ) was reacted with 9,10-phenanthrenequinone (68 μM) in 70% aqueous ethanol containing 0.2 M NaOH at 30°C for 60 min. The reaction mixture was then diluted with an equal volume of 1.2 M HCl and the spectrum recorded. The spectrum was recorded using an emission wavelength of 400- and a 2.5-nm slit width. (From Smith, R. E. and McQuarrie, R., *Anal. Biochem.*, 90, 246, 1978. With permission.)

pounds and various derivatives is reported. The sulfonic acid function provides a basis for the attachment of a "tag" such as norleucine which cajn be used for determining the extent of modification.[14] Reaction with arginine occurs in 0.2 M sodium borate, pH 9.0. Under these conditions, reaction of camphorquinone-sulfonic acid with an amino acid analysis standard showed a greater than 90% loss of arginine and a 25% loss of cystine. Loss of cystine was not observed in the proteins studied (soybean trypsin inhibitor, ribonuclease S-peptide). The arginine derivative is stable for 24 hr in trifluoroacetic acid and under other mild acid conditions. The derivative is stable to 0.5 M hydroxylamine, pH 7.0, conditions under which the cyclohexanedione derivative of arginine decomposes[8] but arginine is regenerated in 0.2 M o-phenylenediamine, pH 8.5 (approximately 75% after 4 hr; complete after 16 hr).

The modification of arginyl residues with hydrazine (aqueous conditions) results in the formation of ornithine but also results in peptide bond cleavage (predominantly at gly-X, X-gly, asn-X, and X-ser peptide bonds).[15]

The determination of the extent of arginine modification is generally determined by amino acid analysis after acid hydrolysis but conditions generally need to be modified to prevent loss of the arginine derivative.[8] This will be discussed for each of the reagents discussed below. The Sakaguchi reaction[16] continues to be useful with recent modifications[17,18] and has recently been used, after acid hydrolysis, to determine the extent of arginine modification by 2,3-butanedione.[19] The use of ninhydrin as a fluorometric reagent for arginine has been described above.[10] Another fluorometric method for the determination of arginine using 9,10-phenanthrenequinone[20] has been described. Figure 5 shows the excitation spectrum for the reaction product of arginine and phenanthrenequinone, while the emission spectrum is shown in Figure 6. The structure of 9,10-phenanthrenequinone is shown in Figure 7 as is the likely reaction with arginine. The time course for the reaction with free arginine is shown in Figure 8 while the time course for the reaction with arginyl residues in proteins is shown in Figure 9. This method is some 1000-fold more sensitive than the Sakaguchi reaction but some concern remains concerning the absolute accuracy of the reagent for determination of arginine in peptide linkage. This is also true of the other reagents.

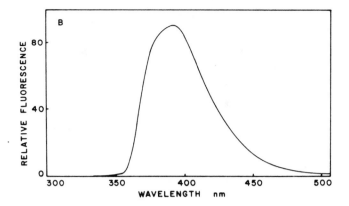

FIGURE 6. The fluorescence emission spectrum of the condensation product of arginine and 9,10-phenanthrenequinone. The reaction conditions are described in Figure 4. The spectrum was recorded using an excitation wavelength of 260 nm and a 2.5-nm slit width. (From Smith, R. E. and McQuarrie, R., *Anal. Biochem.*, 90, 246, 1978. With permission.)

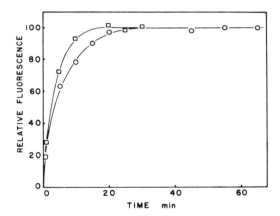

FIGURE 7. The development of fluorescence as a function of the time of reaction of arginine and 9,10-phenanthrenequinone. The reaction was initiated by mixing 1 mℓ of arginine (8 μg) in water with 3 mℓ of 50 μM 9,10-phenanthrenequinone in ethanol and 0.5 mℓ of 2 M NaOH at either 30°C (●) or 44°C (■). At the indicated times, portions were withdrawn and mixed with an equal volume of 1.2 N HCl. The fluorescence was recorded using an excitation wavelength of 312 nm and an emission wavelength of 392 nm. A 5-nm slit width was used. (From Smith, R. E. and McQuarrie, R., *Anal. Biochem.*, 90, 246, 1978. With permission.)

As mentioned above, the use of phenylglyoxal (Figure 10) was developed by Takahashi[2] and has since been applied to the study of the role of arginyl residues in proteins as shown in Table 1. Unfortunately, many of these studies fail to recognize that phenylglyoxal, like glyoxal, will react with α-amino groups at a significant rate[2] (Figure 11). The rate of inactivation of ribonuclease A by phenylglyoxal at different values of pH is shown in Figure 12. Polymerization was noted in a sample incubated for 21 hr. The amino-terminal lysine residue was rapidly modified under these conditions. The possible effect of light on the reaction of phenylglyoxal with arginine as has been reported for 2,3-butanedione[21-23] has not been studied. As noted by Takahashi, the stoichiometry of the reaction involves the

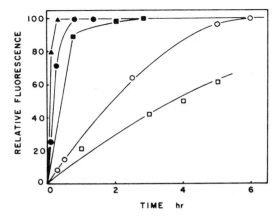

FIGURE 8. The development of fluorescence as a function of time of reaction of proteins with 9,10-phenanthrenequinone. The reaction was initiated by mixing 1 mℓ of protein (100 μg) in 50 mM borate, pH 8.5, with 3 mℓ of 50 μM 9,10-phenanthrenequinone in ethanol and 0.5 mℓ of 2 M NaOH. At the indicated times, portions were removed and mixed with an equal volume of 1.2 N HCl. The fluorescence was recorded using an excitation wavelength of 312 nm and an emission wavelength of 392 nm with a 5 nm-slit width. The reactions were performed either at 30°C (open symbols) or 60°C (closed symbols). The proteins used were bovine serum albumin (□), lysozyme (○), and the β-chain of insulin (△). (From Smith, R. E. and McQuarrie, R., *Anal. Biochem.*, 90, 246, 1978. With permission.)

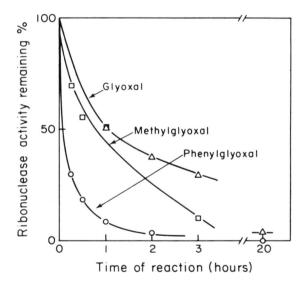

FIGURE 9. The rate of inactivation of ribonuclease A by reaction with phenylglyoxal and related compounds. The reactions were performed in 0.2 M N-ethylmorpholine acetate, pH 8.0, at 25°C. The protein concentration was 0.5% with either 1.5% phenylglyoxal hydrate, 1.5% methylglyoxal, or 1.5% glyoxal hydrate as indicated in the figure. (From Takahashi, K., *J. Biol. Chem.*, 243, 6171, 1968. With permission.)

reaction of 2 mol of phenylglyoxal with 1 mol of arginine (Figure 10). The [14C]-labeled reagent can be easily prepared.[2,24] A facile modification of the original Riley and Gray[24] method which omits the vacuum distillation step has been reported by Hartman and co-

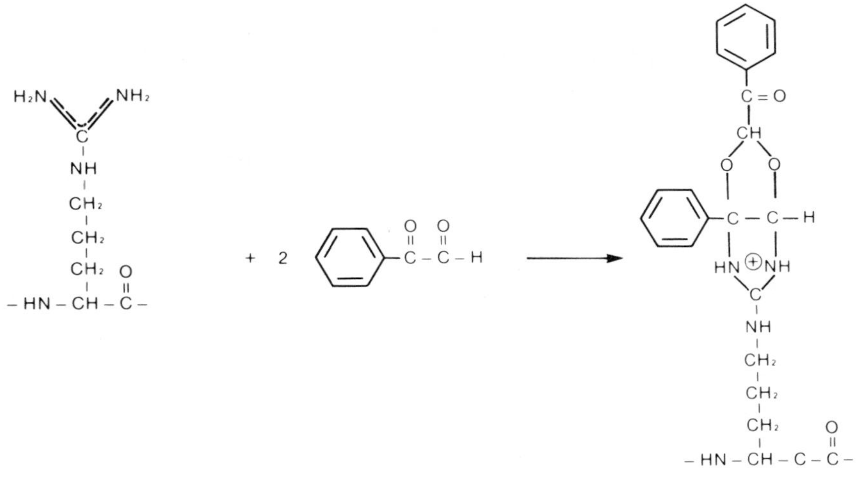

FIGURE 10.   A scheme for the reaction of phenylglyoxal with arginine.

Table 1
REACTION OF PHENYLGLYOXAL WITH ARGINYL RESIDUES IN PEPTIDES AND PROTEINS

Protein	Solvent	Reagent excess[a]	Extent of modification	Ref.
Pancreatic RNase	0.1 M N-ethylmorpholine acetate, pH 8.0	—	2—3/4[b,c]	1
Porcine carboxypeptidase B	0.3 M borate, pH 7.9	200[d]	1/[e]	2
Aspartate transcarbamylase	0.125 M potassium bicarbonate, pH 8.3 or 0.1 M N-ethylmorpholine, pH 8.3	—	2.2/[f,g]	3
Pyruvate kinase	0.1 M triethanolamine, pH 7.0	—	3/28.33[h]	4
Horse liver alcohol dehydrogenase	—	—	—	5, 6
Mitochondrial ATPase	0.097 M sodium borate, 0.097 M EDTA, pH 8.0	—	4/[i]	7
Adenylate kinase	0.1 M triethanolamine ·HCl, pH 7.0	—	/[j]	8, 9
Rhodospirillum rubrum chromatophores	0.05 M borate, pH 8.0	—	/[k]	10
Glutamic acid decarboxylase	0.05 M sodium borate[l]	—	—	11
Ribulosebisphosphate carboxylase	0.066 M sodium[m] bicarbonate, 0.050 M bicine, 0.1 M EDTA, pH 8.0	—	2—3/35[n]	12
Yeast hexokinase	0.035 M Veronal, pH 7.5	—	1/18[o]	13
Propionyl CoA carboxylase	0.050 M borate, pH 8.0	—	—	14
β-Methylcrotonyl CoA carboxylase	0.050 M borate pH 8.0	—	—	14
Superoxide dismutase	0.125 M sodium bicarbonate, pH 8.0	—	1/4[p]	15
Myosin (subfragment 1)	0.1 M potassium bicarbonate, pH 8.0	—	1.7/35[q]	16
Thymidylate synthetase	0.125 M bicarbonate, pH 8.0[r]	—	3.6/12	17

Table 1 (continued)
REACTION OF PHENYLGLYOXAL WITH ARGINYL RESIDUES IN PEPTIDES AND PROTEINS

Protein	Solvent	Reagent excess[a]	Extent of modification	Ref.
Glutamate apodecarboxylase	0.125 *M* sodium` bicarbonate, pH 7.5	—	1/23[t]	18
Adenylate[u] kinase (yeast)	0.025 *M* HEPES, pH 7.5	—	—	19
Cardiac myosin S-1	0.1 *M* N-ethylmorpholine acetate, pH 7.6	—	2.8/42[v]	20
Cystathionase	0.125 *M* bicarbonate, pH 7.9	—	18/45	21
Fatty acid synthetase	0.1 *M* sodium phosphate, 0.0005 *M* dithioerythritol, 0.001 *M* EDTA, pH 7.6	—	4/106	22
Yeast inorganic pyrophosphatase	0.08 *M* N-ethylmorpholine acetate, pH 7.0	—	1/6	23
Porcine phospholipase A	0.125 *M* potassium bicarbonate, pH 8.5	—	1.4/4[w]	24
Superoxide dismutase[x]	0.100 *M* sodium bicarbonate, pH 8.3	50—100	0.88/4.0[y]	25, 26[z]
p-Hydroxybenzoate hydroxylase	0.050 *M* potassium phosphate, pH 8.0	250	2—3/24[aa]	27
Thymidylate synthetase	0.200 *M* N-ethylmorpholine, pH 7.4[bb]	65	2/12[cc]	28
Acetylcholine esterase	0.025 *M* borate, 0.005 phosphate, 0.050 *M* NaCl, pH 7.0	—	3/31[dd]	29
γ-Aminobutyrate aminotransferase	0.05 *M* Tris, pH 8.5	—	—	30
D-β-Hydroxybutyrate dehydrogenase	0.05 *M* HEPES, pH 7.5	—	—[ee]	31
Ornithine transcarboxylase	0.05 *M* Bicine, 0.1 *M* KCl, 0.0001 *M* EDTA, pH 8.05	—	—[ff]	32
Coenzyme B_{12}-dependent diol dehydrase	0.05 *M* borate, pH 8.0	—	—	33
Transketolase	0.125 *M* sodium bicarbonate, pH 7.6	—	4/34[gg]	34
ATP citrate lyase	0.050 *M* HEPES,[hh] pH 8.0	—	8.5/40	36
Malic enzyme	0.037 *M* borate,[ii] pH 7.5	—	—	37
Pyridoxamine-5′-phosphate oxidase	0.1 *M* potassium phosphate, pH 8.0, containing 5% ETOH	—[jj]	6/40	38
Ornithine transcarboxylase	0.125 *M* potassium bicarbonate, pH 8.3	—[kk]	1.5/[ll]	39
Acetate kinase	0.050 *M* triethanolamine, pH 7.6	—[mm]	—[nn]	40
Pancreatic phospholipase A_2	0.2 *M* N-ethylmorpholine, pH 8.0	30	1.0—1.2/[oo]	41
Phosphatidylcholine transfer protein	0.1 *M* sodium bicarbonate, pH 8.0	—	4/10[pp]	42
Aldehyde reductase	0.020 *M* phosphate[qq] pH 7.0	—	0.6/16[rr]	43
Choline acetyltransferase	0.050 *M* HEPES, pH 7.8	—	—[ss]	44
ADP-glucose synthetase	0.05 *M* potassium phosphate, 0.00025 *M* EDTA, pH 7.5	110	1/[tt]	45
Pyruvate oxidase	0.1 *M* sodium phosphate, 0.010 *M* magnesium chloride, pH 7.8	—	2.5/5[uu]	46

Table 1 (continued)
REACTION OF PHENYLGLYOXAL WITH ARGINYL RESIDUES IN PEPTIDES AND PROTEINS

[a] Reagent/protein.

[b] After 3 hr at 25°C.

[c] Had modification of α-amino group and lysine residues.

[d] Reagent/arginine.

[e] After 1 hr at 37°C.

[f] After 3 hr at 25°C.

[g] 1.3/8 In regulatory chain.

[h] 20 Min at 37°C with 23.8 mM phenylglyoxal, protein 1 mg/mℓ.

[i] 30 Min of reaction at 30°C; the presence of efrapeptin, a low-molecular weight antibiotic which is a potent inhibitor of oxidative phosphorylation, prevented the modification of one "fast-reacting" arginyl residue.

[j] A single arginine residue is modified (Arg-97).

[k] A single site appeared to be modified with a second-order rate constant of 1.6 M^{-1} min^{-1}.

[l] pH Not given; reaction at 23°C, kinetic evidence for stoichiometric inactivation.

[m] Solvent made metal-free using BioRad Chelex; reaction performed with and without MgCl$_2$.

[n] Analysis of sulfhydryl groups after phenylglyoxal modification showed no loss of cysteine. These investigations noted the modification with phenylglyoxal is apparently more specific than 2,3-butanedione.

[o] The authors claim 1:1 stoichiometry of phenylglyoxal with the arginyl residue from analysis of dependence of pseudo first-order rate constant vs. reciprocal of reagent (phenylglyoxal concentration). Partial reactivation of modified enzyme was observed reflecting lability of modified arginine residues. Reaction also shows saturation kinetics reflecting "specific" affinity of reagent for enzyme possibly from hydrophobic interaction. These authors suggest that this phenomenon is observed with the reaction of other hydrophobic reagents with this enzyme. A similar phenomenon has been observed with trinitrobenzenesulfonic acid (see Volume I, Chapter 10).

[p] 25°C, 1 hr.

[q] 25°C, 3 mM phenylglyoxal, 3 min.

[r] Rates of enzyme inactivation were dependent upon buffer; at 5.9 mM phenylglyoxal, the following data were obtained, bicarbonate ($T_{1/2}$ = 6.0 min), MOPS ($T_{1/2}$ = 11.5 min), borate ($T_{1/2}$ = 34.0), and phosphate ($T_{1/2}$ = 48.0 min) at 25°C.

[s] These investigators noted a significant buffer effect on the reaction which is more thoroughly explored in Reference 29 of Chapter 1. In this study the following second-order rate constants were obtained with the following reagent/solvent conditions (reactions performed at 23°C): 0.69 M^{-1} min^{-1} with 2,3-butanedione/0.050 M borate, pH 8.0; 33.78 M^{-1} min^{-1} with glyoxal/0.125 M sodium bicarbonate, pH 8.0; 31.00 M^{-1} min^{-1} with methylglyoxal/0.125 M sodium bicarbonate, pH 8.0); and 107.68 M^{-1} min^{-1} with phenylglyoxal/0.125 M sodium bicarbonate, pH 8.0.

[t] 300-Fold excess of reagent, 0.083 M sodium bicarbonate, pH 8.1, 7 min, 23°C.

[u] See more complete discussion of this work in Table 2. 2,3-Butanedione or 1,2-cyclohexanedione appeared to be more effective than phenylglyoxal in this system.

[v] 6 Min, 22°C, 50% loss of activity.

[w] Determined at 99% inactivation (25°C) of phospholipase activity (release of fatty acid from egg yolk in water with 3 mM CaCl$_2$ and 1.4 mM sodium deoxycholate. These investigators (see Reference 24, Table 1) did examine the possibility of amino terminal alanine modification; no loss of alanine was observed with 75% inactivation (0.9 mol Arg modified/mole protein) while enzyme samples with a greater extent of inactivation did have some loss of amino-terminal alanine (quantity not given). These investigators did examine the pH dependence of enzyme inactivation by phenylglyoxal (presumably a direct measure of the rate of arginine modification) and reported the following second-order rate constants (M^{-1} min^{-1}): pH 6.5, 0.3; pH 7.5, 1.5; pH 8.5, 3.3; and pH 9.5, 3.9. These investigators also showed that phenylglyoxal ($T_{1/2}$ = 1 min) was more effective than 2,3-butanedione ($T_{1/2}$ = 20 min) and 1,2-cyclohexanedione ($T_{1/2}$ = 120 min).

[x] Cu, Zn superoxide dismutase from *Saccharomyces cerevisiae*.

[y] Determined at 80% loss of enzymatic activity using reaction of the modified enzyme with 9,10-phenanthrenequinone. This value corresponded to that determined by the incorporation of radiolabeled phenylglyoxal assuming 2:1 adduct. Amino acid analysis with samples prepared using normal hydrolytic conditions (6 N HCl, 110°C, 20 hr) suggested only approximately 50% of this extent of arginine modification. When thioglycolic acid was included during the hydrolysis, values for the extent of arginine modification approached those determined by the fluorescence technique and radiolabel incorporation.

[z] The study is an extension of the observation reported in Reference 25 and uses reaction with 4-hydroxy-3-nitrophenylglyoxal, a chromophoric derivative of phenylglyoxal, to identify the specific arginine residue modified. It is of some interest that the rate of reaction with this derivative is approximately sixfold less than that with the parent phenylglyoxal.

Table 1 (continued)
REACTION OF PHENYLGLYOXAL WITH ARGINYL RESIDUES IN PEPTIDES AND PROTEINS

[aa] Reaction at 25°C for 60 to 120 min. Loss of lysine residues was not observed under these reaction conditions. Amino acid analysis (hydrolysis in 6 N HCl, 110°C, 24 hr) correlated well with radiolabeled phenylglyoxal incorporation assuming 2:1 stoichiometry (i.e., amino acid analysis gave 3.6 mol Arg lost/mole enzyme while 7.54 mol radiolabel was incorporated).

[bb] These investigators (see Reference 28, Table 1) examined the reaction at pH 7.4 (rate of inactivation of 32 M^{-1} min^{-1}). An approximate 100-fold increase in the rate of inactivation.

[cc] The presence of substrate, 2'-deoxyuridylate, prevents the modification of 1 mol of arginine per mole of enzyme. It is noted that these results differ from those reported in Reference 17. There were differences in solvent conditions. It is not clear why this would account for the differences observed in these two studies. It is noted that the investigators in Reference 17 obtained similar stoichiometry with 2,3-butanedione.

[dd] The modification with phenylglyoxal is associated with an approximate 15% loss of enzyme activity. Treatment with 2,3-butanedione under similar reaction conditions results in the modification of approximately one more mole of arginine per mole enzyme with an approximate 75% loss of catalytic activity.

[ee] Stoichiometry was not established but the data are consistent with the loss of activity resulting from the modification of a single arginine residue. Submitochondrial vesicles were used as the source of enzyme in these studies. A second-order rate constant of 1.03 M^{-1} min^{-1} was obtained from the reaction with phenylglyoxal. A value of 0.8 M^{-1} min^{-1} was obtained for reaction with 1,2-cyclohexanedione (0.050 M borate, pH 7.5) while a value of 4.6 M^{-1} min^{-1} was obtained for 2,3-butanedione in the borate buffer system.

[ff] See more complete discussion of this study under Table 2. For inactivation by phenylglyoxal, a second-order rate constant of 56 M^{-1} min^{-1} was obtained at pH 8.04. The reactions were performed in the dark.

[gg] Analysis of Tsou plots[35] indicates at least two classes of residues react at different rates.

[hh] Most studies were performed in this solvent at 30°C with a second-order rate constant of 0.33 M^{-1} sec^{-1}. The rate was reduced in potassium phosphate (k = 0.25 M^{-1} sec^{-1} and borate (k = 0.078 M^{-1} sec^{-1}).

[ii] Under these conditions at 24°C, a second-order rate constant of k = 7.08 M^{-1} min^{-1} assuming that the rate of inactivation is directly related to the modification of arginine. With 2,3-butanedione in 0.048 M borate a second-order rate constant of k = 5.4 M^{-1} min^{-1} is compared to 1.69 M^{-1} min^{-1} with methylglyoxal and 0.032 M^{-1} min^{-1} with 2,4-pentanedione.

[jj] The rate of inactivation at 25°C for the apoenzyme was determined to be 3.7 M^{-1} min^{-1} and 11.1 M^{-1} min^{-1} for the holoenzyme.

[kk] A second-order rate constant of k = 4.6 M^{-1} min^{-1} at 25°C was obtained under these conditions.

[ll] Based on incorporation of radiolabeled phenylglyoxal, 1.5 arginine residues are modified per 35,000 chain after 3 hr of reaction. There are likely different classes of reactive arginyl residues where the more reactive group(s) directly associated with catalytic activity.

[mm] Saturation kinetics are observed with phenylglyoxal suggesting the formation of an enzyme-inhibitor complex prior to reaction with an arginine residue(s).

[nn] With 95% loss of catalytic activity there is 94% modification of arginine.

[oo] See Reference 24 for somewhat differing results. This study shows that this level of arginine modification is associated with 80% loss of amino-terminal alanine. It was necessary to protect the α-amino group of the amino-terminal alanine with a *t*-butyloxycarbonyl group to avoid modification under these reaction conditions. The use of radiolabeled cyclohexanedione established Arg-6 as the primary site of modification.

[pp] 30 Min at 25°C. Extent of modification based on radiolabel incorporation and amino acid analysis.

[qq] For reaction at 30°C, a second-order rate constant of k = 2.6 M^{-1} min^{-1} assuming that the loss of activity seen with phenylglyoxal directly reflected the loss of an arginine residue(s).

[rr] Determined from both amino acid analysis and radiolabel incorporation.

[ss] Phenylglyoxal was much more effective than 2,3-butanedione or camphorquinone-10-sulfonic acid.

[tt] Assuming 2:1 stoichiometry of phenylglyoxal to arginine; reaction at 25°C. Phenglyglyoxal is much more effective than 1,2-cyclohexanedione (twofold molar excess of 1,2-cyclohexanedione had $T_{1/2}$ = 24 min).

[uu] From radiolabel incorporation assuming 2:1 stoichiometry. There are clearly at least two classes of reactive arginine residue. When the reaction is performed at pH 6.0, inactivation with phenylglyoxal can be partially reversed on dilution in pH 6.0 buffer.

References for Table 1

1. **Takahashi, K.,** The reaction of phenylglyoxal with arginine residues in proteins, *J. Biol. Chem.*, 243, 6171, 1968.

Table 1 (continued)

2. **Werber, M. M. and Sokolovsky, M.,** Chemical evidence for a functional arginine residue in carboxypeptidase B, *Biochem. Biophys. Res. Commun.,* 48, 384, 1972.
3. **Kantrowitz, E. R. and Lipscomb, W. N.,** Functionally important arginine residues of aspartate transcarbamylase, *J. Biol. Chem.,* 252, 2873, 1977.
4. **Berghäuser, J.,** Modifizierung von argininresten in pyruvat-kinase, *Hoppe-Seyler's Physiol. Chem.,* 358, 1565, 1977.
5. **Lange, L. G., III, Riordan, J. F., and Vallee, B. L.,** Functional argininyl residues as NADH binding sites of alcohol dehydrogenases, *Biochemistry,* 13, 4361, 1974.
6. **Jörnvall, H., Lange, L. G., III, Riordan, J. F., and Vallee, B. L.,** Identification of a reactive arginyl residue in horse liver alcohol dehydrogenase, *Biochem. Biophys. Res. Commun.,* 77, 73, 1977.
7. **Kohlbrenner, W. E. and Cross, R. L.,** Efrapeptin prevents modification by phenylglyoxal of an essential arginyl residue in mitochondrial adenosine triphosphatase, *J. Biol. Chem.,* 253, 7609, 1978.
8. **Berghäuser, J.,** A reactive arginine in adenylate kinase, *Biochim. Biophys. Acta,* 397, 370, 1975.
9. **Berghäuser, J. and Schirmer, R. H.,** Properties of adenylate kinase after modification of Arg-97 by phenylglyoxal, *Biochim. Biophys. Acta,* 537, 428, 1978.
10. **Vallejos, R. H., Lescano, W. I. M., and Lucero, H. A.,** Involvement of an essential arginyl residue in the coupling activity of *Rhodospirillum rubrum* chromatophores, *Arch. Biochem. Biophys.,* 190, 578, 1978.
11. **Tunnicliff, G. and Ngo, T. T.,** Functional role of arginine residues in glutamic acid decarboxylase from brain and bacteria, *Experientia,* 34, 989, 1978.
12. **Schloss, J. V., Norton, I. L., Stringer, C. D., and Hartman, F. C.,** Inactivation of ribulosebisphosphate carboxylase by modification of arginyl residues with phenylgloxal, *Biochemistry,* 17, 5626, 1978.
13. **Philips, M., Pho, D. B., and Pradel, L.-A.,** An essential arginyl residue in yeast hexokinase, *Biochim. Biophys. Acta,* 566, 296, 1979.
14. **Wolf, B., Kalousek, F., and Rosenberg, L. E.,** Essential arginine residues in the active sites of propionyl CoA carboxylase and beta-methylcrotonyl CoA carboxylase, *Enzyme,* 24, 302, 1979.
15. **Malinowski, D. P. and Fridovich, I.,** Chemical modification of arginine at the active site of the bovine erythrocyte superoxide dismutase, *Biochemistry,* 18, 5909, 1979.
16. **Mornet, D., Pantel, P., Audemard, E., and Kassab, R.,** Involvement of an arginyl residue in the catalytic activity of myosin heads, *Eur. J. Biochem.,* 100, 421, 1979.
17. **Cipollo, K. L. and Dunlap, R. B.,** Essential arginyl residues in thymidylate synthetase from amethopterin-resistant *Lactobacillus casei, Biochemistry,* 18, 5537, 1979.
18. **Cheung, S.-T. and Fonda, M. L.,** Kinetics of the inactivation of *Escherichia coli* glutamate apodecarboxylase by phenylglyoxal, *Arch. Biochem. Biophys.,* 198, 541, 1979.
19. **Varimo, K. and Londesborough, J.,** Evidence for essential arginine in yeast adenylate cyclase, *FEBS Lett.,* 106, 153, 1979.
20. **Morkin, E., Flink, I. L., and Banerjee, S. K.,** Phenylglyoxal modification of cardiac myosin S-1. Evidence for essential arginine residues at the active site, *J. Biol. Chem.,* 254, 12647, 1979.
21. **Portemer, C., Pierre, Y., Loriette, C., and Chatagner, F.,** Number of arginine residues in the substrate binding sites of rat liver cystathionase, *FEBS Lett.,* 108, 419, 1979.
22. **Poulose, A. J. and Kolattukudy, P. E.,** Presence of one essential arginine that specifically binds the 2'-phosphate of NADPH on each of the ketoacyl reductase and enoyl reductase active sites of fatty acid synthetase, *Arch. Biochem. Biophys.,* 199, 457, 1980.
23. **Bond, M. W., Chiu, N. Y., and Cooperman, B. S.,** Identification of an arginine residue important for enzymatic activity within the covalent structure of yeast inorganic pyrophosphatase, *Biochemistry,* 19, 94, 1980.
24. **Vensel, L. A. and Kantrowitz, E. R.,** An essential arginine residue in porcine phospholipase A_2, *J. Biol. Chem.,* 255, 7306, 1980.
25. **Borders, C. L., Jr. and Johansen, J. T.,** Essential arginyl residues in Cu, Zn superoxide dismutase from *Saccharomyces cerevisiae, Carlsberg Res. Commun.,* 45, 185, 1980.
26. **Borders, C. L., Jr. and Johansen, J. T.,** Identification of Arg-143 as the essential arginyl residue in yeast Cu, Zn superoxide dismutase by the use of a chromophoric arginine reagent, *Biochem. Biophys. Res. Commun.,* 96, 1071, 1980.
27. **Shoun, H., Beppu, T., and Arima, K.,** An essential arginine residue at the substrate-binding site of *p*-hydroxybenzoate hydroxylase, *J. Biol. Chem.,* 255, 9319, 1980.
28. **Belfort, M., Maley, G. F., and Maley, F.,** A single functional arginyl residue involved in the catalysis promoted by *Lactobacillus casei* thymidylate synthetase, *Arch. Biochem. Biophys.,* 204, 340, 1980.
29. **Müllner, H. and Sund, H.,** Essential arginine residue in acetylcholinesterase from *Torpedo californica, FEBS Lett.,* 119, 283, 1980.
30. **Tunnicliff, G.,** Essential arginine residues at the pyridoxal phosphate binding site of brain α-aminobutyrate aminotransferase, *Biochem. Biophys. Res. Commun.,* 97, 160, 1980.

Table 1 (continued)

31. **El Kebbaj, M. S., Latruffe, N., and Gaudemer, Y.,** Presence of an essential arginine residue in D-β-hydroxybutyrate dehydrogenase from mitochondrial inner membrane, *Biochem. Biophys. Res. Commun.,* 96, 1569, 1980.
32. **Marshall, M. and Cohen, P. P.,** Evidence for an exceptionally reactive arginyl residue at the binding site for carbamyl phosphate in bovine ornithine transcarbamylase, *J. Biol. Chem.,* 255, 7301, 1980.
33. **Kuno, S., Toraya, T., and Fukui, S.,** Coenzyme B$_{12}$-dependent diol dehydrase: chemical modification with 2,3-butanedione and phenylglyoxal, *Arch. Biochem. Biophys.,* 205, 240, 1980.
34. **Kremer, A. B., Egan, R. M., and Sable, H. Z.,** The active site of transketolase. Two arginine residues are essential for activity, *J. Biol. Chem.,* 255, 2405, 1980.
35. **Tsou, C.-L.,** Relation between modification of functional groups of proteins and their biological activity. I. A graphical method for the determination of the number and type of essential groups, *Sci. Sin.,* 11, 1535, 1962.
36. **Ramakrishna, S. and Benjamin, W. B.,** Evidence for an essential arginine residue at the active site of ATP citrate lyase from rat liver, *Biochem. J.,* 195, 735, 1981.
37. **Chang, G.-G. and Huang, T.-M.,** Modification of essential arginine residues of pigeon liver malic enzyme, *Biochim. Biophys. Acta,* 660, 341, 1981.
38. **Choi, J.-D. and McCormick, D. B.,** Roles of arginyl residues in pyridoxamine-5'-phosphate oxidase from rabbit liver, *Biochemistry,* 20, 5722, 1981.
39. **Fortin, A. F., Hauber, J. M., and Kantrowitz, E. R.,** Comparison of the essential arginine residue in *Escherichia coli* ornithine and aspartate transcarbamylases, *Biochim. Biophys. Acta,* 662, 8, 1981.
40. **Wong, S. S. and Wong, L.-J.,** Evidence for an essential arginine residue at the active site of *Escherichia coli* acetate kinase, *Biochim. Biophys. Acta,* 660, 142, 1981.
41. **Fleer, E. A. M., Puijk, W. C., Slotboom, A. J., and DeHaas, G. H.,** Modification of arginine residues in porcine pancreatic phospholipase A$_2$, *Eur. J. Biochem.,* 116, 277, 1981.
42. **Akeroyd, R., Lange, L. G., Westerman, J., and Wirtz, K. W. A.,** Modification of the phosphatidylcholine-transfer protein from bovine liver with butanedione and phenylglyoxal. Evidence for one essential arginine residue, *Eur. J. Biochem.,* 121, 77, 1981.
43. **Branlant, G., Tritsch, D., and Biellmann, J.-F.,** Evidence for the presence of anion-recognition sites in pig-liver aldehyde reductase. Modification by phenylglyoxal and *p*-carboxyphenyl glyoxal of an arginyl residue located close to the substrate-binding site, *Eur. J. Biochem.,* 116, 505, 1981.
44. **Mautner, H. G., Pakyla, A. A., and Merrill, R. E.,** Evidence for presence of an arginine residue in the coenzyme A binding site of choline acetyltransferase, *Proc. Nat. Acad. Sci. U.S.A.,* 78, 7449, 1981.
45. **Carlson, C. A. and Preiss, J.,** Involvement of arginine residues in the allosteric activation of *Escherichia coli* ADP–glucose synthetase, *Biochemistry,* 21, 1929, 1982.
46. **Koland, J. G., O'Brien, T. A., and Gennis, R. B.,** Role of arginine in the binding of thiamin pyrophosphate to *Escherichia coli* pyruvate oxidase, *Biochemistry,* 21, 2656, 1982.

workers.[25] Radiolabeled acetophenone was added to an equal amount (on the basis of weight) of selenium dioxide in dioxane-water (30:1). The mixture was refluxed for 3 hr after which solvent was removed under a stream of nitrogen. The residue was taken up in boiling water and activated charcoal added. The hot slurry was filtered through Celite. The phenylglyoxal crystallized spontaneously from the filtrate on cooling.

The synthesis of phenyl [2-³H] glyoxal[26] has been reported. Borders and co-workers[27] have reported the synthesis of a chromophoric derivative, 4-hydroxy-3-nitrophenylglyoxal, which should prove quite useful in the study of arginyl residues. *p*-Hydroxylphenylglyoxal[28] has also been used as a spectrophotometric reagent for the study of this reaction.

Of particular interest has been the observations of Fonda and Cheung[29] that the reaction of arginine with phenylglyoxal is greatly accelerated in bicarbonate-carbonate buffer systems. Figure 13 shows the reaction of phenylglyoxal with *N*-acetylarginine, *N*-acetyllysine and *N*-acetylcysteine in 0.083 *M* sodium bicarbonate, pH 7.5. Reaction is only seen, for all practical purposes, with the arginine derivative. L-Arginine reacted in the same manner suggesting that modification of the α-amino group did not occur under these conditions. Figure 14 compares the rate of reaction of phenylglyoxal with arginine in bicarbonate buffer with that in other buffer systems (borate, Veronal, *N*-ethylmorpholine). The reaction appears to be first order with respect to bicarbonate (Figure 15). The reaction of methylglyoxal with arginine is also enhanced by bicarbonate (Figure 16) while a similar effect is not seen with

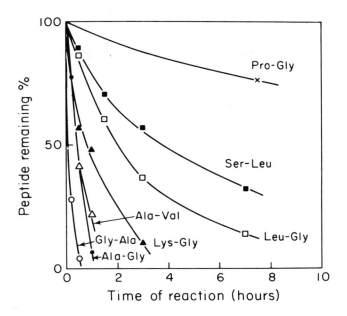

FIGURE 11.   The rate of reaction of phenylglyoxal with various di-
peptides. The reactions were performed in 0.2 *M* *N*–ethylmorpholine
acetate, pH 8.0, at 25°C. The concentration of peptide was 0.017%
and the concentration of phenylglyoxal hydrate was 1.25%. At the
indicated times a 50-μℓ portion was withdrawn from the reaction mixture, diluted into 1.2 mℓ sodium citrate, pH 2.2, and stored at − 10°C
until analysis for residual peptide on the amino acid analyzer and for
amino acids after acid hydrolysis. (From Takahashi, K., *J. Biol. Chem.*,
243, 6171, 1968. With permission.)

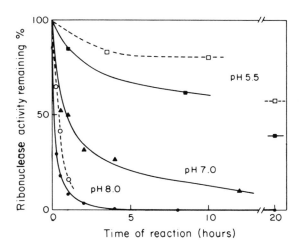

FIGURE 12.   The rate of inactivation of bovine pancreatic ribonuclease A by phenylglyoxal
as a function of pH. The concentration of protein was 0.5% and the concentration of phenylglyoxal hydrate was 1.5% at 25°C under the following conditions: ○-----○, pH 8.0 (0.1 *M*
N-ethylmorpholine acetate; ●———●, pH 8.0 (0.1 *M* *N*-ethylmorpholine acetate with 0.6%
2′(3′)-cytidylic acid); ▲———▲, pH 7.0 (0.1 *M* sodium phosphate); □-----□, pH 5.5 (0.1 *M*
sodium acetate); ■———■, pH 5.5 (0.1 *M* sodium acetate with 0.6% 2′(3′)-cytidylic acid.
(From Takahashi, K., *J. Biol. Chem.*, 243, 6171, 1968. With permission.)

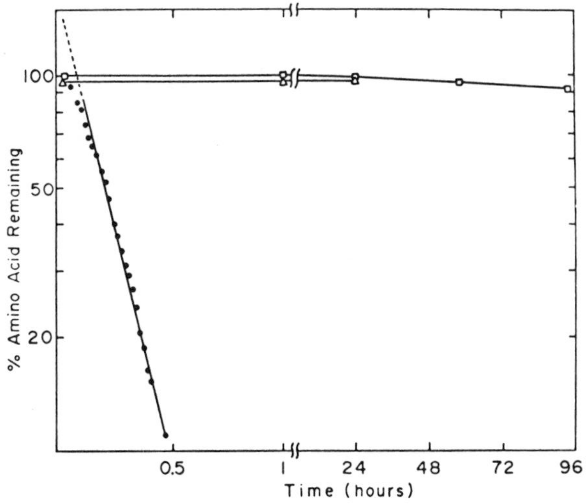

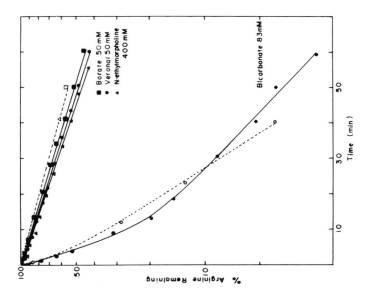

FIGURE 14. The reaction of arginine with phenylglyoxal in various buffers. The reactions mixtures contained 5 m*M* L-arginine and 25 m*M* phenylglyoxal in the designated buffer at pH 7.5 and 25°C. Portions were removed at the indicated times, and the arginine concentrations were determined by the Sakaguchi test (solid lines and closed symbols) or by amino acid analysis (dashed lines and open symbols). (From Cheung, S.-T. and Fonda, M. L., *Biochem. Biophys. Res. Commun.*, 90, 940, 1979. With permission.)

FIGURE 13. The modification of amino acid derivatives with phenylglyoxal in bicarbonate buffer. Shown is the time course for the reaction of *N*-acetylarginine (●), *N*-acetyllysine (□), and *N*-acetylcysteine (△) with phenylglyoxal in 83 m*M* bicarbonate buffer, pH 7.5, at 23°C. The reaction with *N*-acetylarginine was monitored by the increase in absorbance at 340 nm. The amounts of unreacted *N*-acetyllysine and *N*-acetylcysteine were determined by reaction with 2,4,6-trinitrobenzenesulfonic acid. (From Cheung, S.-T. and Fonda, M. L., *Biochem. Biophys. Res. Commun.*, 90, 940, 1979. With permission.)

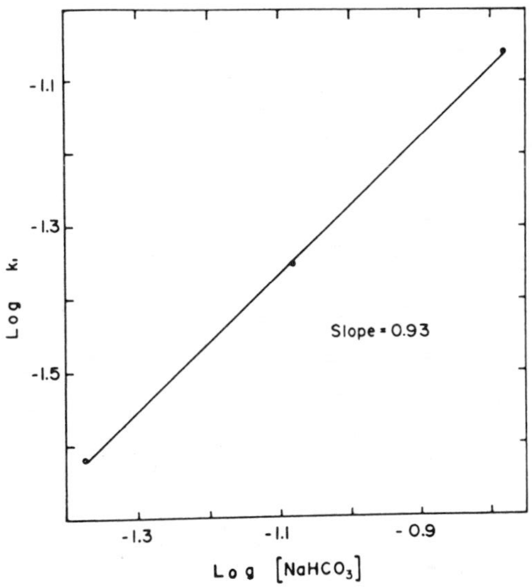

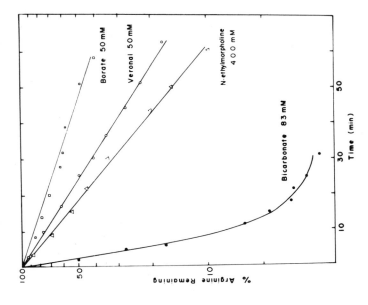

FIGURE 16.   The effect of various buffers on the rate of reaction of arginine with methylglyoxal. The reaction mixtures contained 5 m*M* arginine and 25 m*M* methylglyoxal in the buffers indicated. Portions were removed at the indicated times for the determination of the amount of arginine remaining by the Sakaguchi test. (From Cheung, S.-T. and Fonda, M. L., *Biochem. Biophys. Res. Commun.*, 90, 940, 1979. With permission.)

FIGURE 15.   The effect of bicarbonate concentration on the rate of reaction of arginine with phenylglyoxal. Shown is a plot of logarithm apparent first-order rate constants vs. logarithm bicarbonate concentrations. The reaction mixtures contained 5 m*M N*-acetylarginine and 25 m*M* phenylglyoxal in sodium bicarbonate, pH 7.5, at 25°C. The absorbance at 340 nm was recorded, and the rate constants were obtained from the slopes of the plots of ln $(A_\infty - A_t)$ vs. time. The slope of the line obtained in this figure is 0.93 suggesting that the reaction is first-order with respect to bicarbonate. (From Cheung, S.-T. and Fonda, M. L., *Biochem. Biophys. Res. Commun.*, 90, 940, 1979. With permission.)

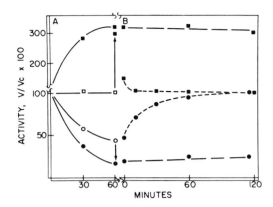

FIGURE 17.   The modification of carboxypeptidase A with 2,3-butanedione in borate buffer. (A) Changes in esterase (□, ■) and peptidase (○, ●) activities on modification of carboxy-peptidase A (0.15 mM) with 2,3-butanedione in 0.05 M borate — 1 M NaCl, pH 7.5 (9 mM reagent, closed symbols), or in 0.02 M Veronal — 1.0 M NaCl, pH 7.5 (75 mM reagent, open symbols) at 20°C. The changes in the activity immediately on addition of borate after 1 hr to the sample reacted in Veronal buffer are indicated by the arrows. (B) Changes in activities of the samples reacted in borate buffer subsequent to gel filtration through Bio-Gel P-4 equilibrated either with 0.05 M borate — 1.0 M NaCl, pH 7.5 (— ——), or with 0.02 M Veronal — 1.0 M NaCl, pH 7.5 (----). (From Riordan, J. F., *Biochemistry*, 12, 3915, 1973. With permission.)

either glyoxal or 2,3-butanedione. The molecular basis for this specific buffer effect is not clear at this time nor is it known whether reaction with α-amino functional groups occurs at a different rate than with other solvent systems used for this modification of arginine with phenylglyoxal. Feeney and co-workers[30] reported that *p*-nitrophenylglyoxal (prepared from *p*-nitroacetophenone — see Reference 31) reacts with arginine in 0.17 sodium pyrophosphate — 0.15 M sodium ascorbate, pH 9.0 to yield a derivative which absorbs at 475 nm. There is also reaction with histidine (the imidazole ring is critical for this reaction in that the 1-methyl derivative yielded a derivative which absorbed at 475 nm while the 3-methyl derivative did not). Free sulfhydryl groups also yielded a product with absorbance at 475 nm, but its absorbance was only 3% of that of the arginine. Branlant and co-workers[32] have used *p*-carboxyphenyl glyoxal in bicarbonate buffer at pH 8.0 to modify aldehyde reductase. Saturation kinetics were noted with the use of this reagent.

The synthesis of a heterobifunctional reagent derived from phenylglyoxal, *p*-azido-phen-ylglyoxal, has been reported.[33]

2,3-Butanedione is a second well-characterized reagent for the selective modification of arginyl residues in proteins. Yankeelov and co-workers introduced the use of this re-agent.[3,34,35] There were problems with the specificity of the reaction (c.f. Reference 35) and the time required for modification until the observation of Riordan[36] that borate had a significant effect on the nature of the reaction of 2,3-butanedione with arginyl residues in proteins. Figure 17 shows the effect of borate (0.05 M borate, 1.0 M NaCl, pH 7.5) on the changes in biological activity occurring on the reaction of carboxypeptidase A (0.15 mM) with 2,3-butanedione (freshly distilled). Note that in particular, the enhancement of esterase activity in the presence of butanedione is dependent on the presence of borate buffer as no significant change is seen with butanedione in 0.02 M Veronal, 1.0 M NaCl, pH 7.5. The removal of borate by gel filtration results in the recovery of activity.

The ability of 2,3-butanedione to act as a photosensitizing agent for the destruction of amino acids and proteins in the presence of oxygen was emphasized in work by Fliss and

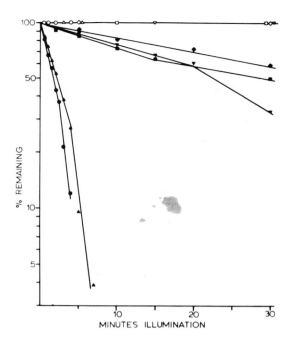

FIGURE 18.   2,3-Butanedione-sensitized destruction of α-amino acids. ●, tryptophan (99.5 μM); ■, tyrosine (333 μM); ▲, histidine (92 μM); ▼, methionine (1000 μM); and ◆, cystine (33.3 μM) in the presence of 2,3-butanedione (9580 μM) and continuous oxygenation were irradiated at pH 6.0 (irradiation was performed in quartz cuvettes 20 cm from a "Blak-Lite" UV light source, Canlab catalog No. L6093-1, equipped with a 100 W lamp emitting light almost exclusively in the range of 350 to 375 nm) at 36°C. Open symbols represent preparations of amino acids (at the same concentrations as the experiments described above) irradiated in the absence of 2,3-butanedione. The above experiments used freshly distilled monomer preparation of 2,3-butanedione. (From Fliss, H. and Viswanatha, T., *Can. J. Biochem.*, 57, 1267, 1979. With permission.)

Viswanatha.[21] Figure 18 shows the destruction of certain amino acids in the presence of 2,3-butanedione and oxygen at pH 6.0 (phosphate) 36°C upon irradiation at 350 to 375 nm ("Blak-Lite" UV-Lamp, 100 W bulb, 20 cm from sample contained in a quartz cuvette). As would be expected from consideration of early photooxidation work, tryptophan and histidine are lost most rapidly with methionine; cystine and tyrosine are lost at a much slower rate. Loss is not seen on irradiation in the absence of 2,3-butanedione (open symbols). Azide (10 mM), a singlet oxygen scavenger, greatly reduced the rate of loss of amino acids. The absence of oxygen also greatly reduces the rate of loss of sensitive amino acids. Figure 19 shows a similar experiment with α-chymotrypsin. Note the rapid loss of tryptophan in the sample irradiated in the presence of 2,3-butanedione. Again the presence of azide and the absence of oxygen reduced the extent of inactivation and tryptophan modification.

These observations have been confirmed and extended by other laboratories.[22,23] An examination of recent studies using 2,3-butanedione to modify arginyl residues in proteins is presented in Table 2.

The use of 1,2-cyclohexanedione under very basic conditions to modify arginyl residues was demonstrated in 1967.[37] However, it was not until Patthy and Smith[8] reported on the reaction of 1,2-cyclohexanedione in borate with arginyl residues in proteins that the use of this reagent became practical. These investigators reported that 1,2-cyclohexanedione reacted with arginyl residues in 0.2 M borate, pH 9.0. At alkaline pH, reaction of 1,2-cyclohexa-

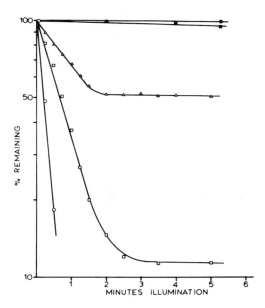

FIGURE 19.   The rate of photodynamic destruction of enzymatic activity and tryptophan in irradiated α-chymotrypsin. Portions (3 mℓ) of α-chymotrypsin (6.7 μM) at pH 4.0 (0.1 M acetate) were irradiated (see caption to Figure 18) in the presence (open symbols) or absence (closed symbols) of 0.1 M 2,3-butanedione at 36°C with continuous oxygenation. The circles represent esterase activity and the squares represent tryptophan (determined by titration with *N*-bromosuccinimide). The open triangles represent the absorbance at 280 nm of preparations irradiated in the presence of 2,3-butanedione. (From Fliss, H. and Viswanatha, T., *Can. J. Biochem.*, 57, 1267, 1979. With permission.)

Table 2
USE OF 2,3-BUTANEDIONE TO MODIFY ARGINYL RESIDUES IN PROTEINS

Protein	Solvent	Reagent excess[a]	Stoichiometry	Ref.
Carboxypeptidase A	0.05 M borate, 1.0 M NaCl, pH 7.5	—[b]	2/10	1
Chymotrypsin	0.1 M phosphate, pH 6.0	100[c]	1/3[d]	2
Thymidylate synthetase	0.050 M borate, pH 8.0	—	—[e]	3
Prostatic acid phosphatase	0.050 M borate, pH 8.0	—	—[f]	4
Purine nucleoside phosphorylase	0.0165 M borate, pH 8.0	—	—[g]	5
Yeast hexokinase PII	0.050 M borate, pH 8.3	—	4.2/18[h]	6
Isocitrate dehydrogenase	0.05 M MES, pH 6.2, 20% glycerol, 0.0021 M MnSO$_4$	—	1.6/13.4[i]	7
Stearylcoenzyme A desaturase	0.050 M sodium borate, pH 8.1	2500	2/[j]	8
Superoxide dismutase	0.050 M borate, pH 9.0	—[k]	1.3/4[l]	9
Energy-independent transhydrogenase	0.050 M sodium[m] borate, pH 7.8	—	—	10
Enolase	0.050 M borate, pH 8.3, 0.001 M Mg (OAc)$_2$, 0.01 mM EDTA	260	3/16[n]	11
NADPH-dependent aldehyde reductase	0.050 M borate, pH 7.0	—[o]	1/18[p]	12
Aryl sulfatase A	0.050 M[q] NaHCO$_3$, pH 8.0	—	—	13
Na$^+$, K$^+$-ATPase	0.04 M TES, 0.02 M borate, pH 7.4	—	—	14

Table 2 (continued)
USE OF 2,3-BUTANEDIONE TO MODIFY ARGINYL RESIDUES IN PROTEINS

Protein	Solvent	Reagent excess[a]	Stoichiometry	Ref.
Carbamate kinase	0.005 M triethanolamine, 0.050 M borate, pH 7.5	2000	1.2/3.0[r]	15
Thymidylate synthetase	0.050 M borate, 0.001 M EDTA, pH 8.0	1201	2.1/12[s]	16
$(K^+ + H^+)$-ATPase	0.125 M sodium borate, pH 7.0	—	—[t]	17
Cu, Zn superoxide dismutase	0.050 M borate, pH 8.3	—[u]	—	18
Fatty acid synthetase	0.020 M borate, 0.200 M KCl, 0.001 M dithiothreitol, 0.001 mM EDTA, pH 7.6	—	—[v]	19
Acetylcholinesterase	0.005 M phosphate, 0.025 M borate, 0.050 M NaCl, pH 7.0	—	4/31[w]	20
Coenzyme B_{12}-dependent diol dehydrase	0.050 M borate, pH 8.5	—	—[x]	21
Ornithine transcarbamylase	0.05 M bicine,[y] 0.1 mM EDTA, 0.1 M KCl, pH 7.67	—	0.88/11[z]	22
Glycogen phosphorylase	0.020 M sodium tetraborate, 1 mM EDTA, pH 7.5	—	—	23
Cytochrome c	0.05 M sodium bicarborate, pH 7.5	9900[aa]	2/2[bb]	24
Bacteriorhodopsin	0.100 M borate, pH 8.2	66,700	4/79[cc]	25
α-Ketoglutarate dehydrogenase	0.050 M sodium borate, pH 8.0	—[dd]	—[ee]	26
Acetate kinase	0.050 M borate, pH 8.6	—	—	27
Malic enzyme	0.045 M borate,[ff] pH 7.5	—	—	28
Glucose phosphate isomerase	0.05 M sodium borate, pH 8.7	—	7.8/30[gg]	29
Saccharopine dehydrogenase	0.08 M HEPES, 0.2 M KCl, 0.01 M borate, pH 8.0	—[hh]	8/38[ii]	30
Testicular hyaluronidase	0.050 M borate, pH 8.3	—[jj]	3.6/28	31
Glutathione reductase	0.050 M sodium borate, pH 8.3, 1 mM EDTA	20,000	5.3/[kk]	32

[a] Mole reagent per mole protein unless otherwise indicated.
[b] This study demonstrated that, in the presence of borate, there is essentially no difference in the reaction of 2,3-butanedione monomer and butadione trimer. It is noted that the commercially available 2,3-butanedione should be distilled immediately prior to use.
[c] This study used 2,3-butanedione trimer prepared by allowing 2,3-butanedione (40 mℓ) to stand with 80 g untreated Permutit under dry air (after shaking to obtain an even dispension of 2,3-butanedione in Permutit) for 4 to 6 weeks at ambient temperature. The mixture was extracted with anhydrous ether. The ether extract was taken to an oil with dry air. The oil was allowed to stand for 5 to 7 days to permit crystallization of the timer.
[d] In the absence of light, also some loss of lysine; no loss of catalytic activity. In the presence of sunlight there was rapid inactivation of the enzyme with loss of lysine, arginine (less than in the dark), and tyrosine. With the exception of tyrosine modification, the changes in amino acid composition in the reaction exposed to light were less than those for the dark reaction despite the more significant loss of activity. Study of the wavelength dependence demonstrates that light of 300 nm is most effective. 2,3-Butanedione monomer was not effective in this photoinactivation process.
[e] Stoichiometry of reaction not established. Inactivation was reversed by gel filtration in 0.05 M Tris, 0.010 M β-mercaptoethanol, pH 8.0.
[f] 30°C.
[g] Ambient temperature. Calf spleen enzyme had 26 Arg modified at 98% loss of activity. Reaction with arginyl residues (as judged by loss of catalytic activity) was 50% as rapid with 2,3-butanedione in borate ($T_{1/2}$ = 40.3 min) as with phenylglyoxal in Tris buffer ($T_{1/2}$ = 19.2 min).

Table 2 (continued)
USE OF 2,3-BUTANEDIONE TO MODIFY ARGINYL RESIDUES IN PROTEINS

h   Reaction at 25°C. Determined by amino acid analysis after acid hydrolysis (6 N HCl, 110°C, 18 hr). MgATP (5 mM) did not protect against either modification or loss of enzymatic activity but MgATP and glucose reduced extent of modification from 3.3 arginine residues per subunit (65% inactivation) to 2.1 residues per subunit (20% inactivation). Inactivation was also observed with phenylglyoxal in 0.050 M BICINE, pH 8.3. Stoichiometry with this modification was not established.

i   Determined by amino acid analysis. As indicated, the maximum value obtained is 1.6 residues modified out of an average of 13.4 arginyl residues per subunit.

j   The modification was performed at 25°C. The presence of stearyl-CoA greatly decreased the rate and extent of inactivation by 2,3-butanedione. When the modified enzyme is taken into 0.020 Tris (acetate), 0.100 M NaCl, pH 8.1 by gel filtration there is the rapid recovery of activity and the concomitant decrease in the extent of arginine modification. A similar extent of modification and loss of catalytic activity was seen with 1,2-cyclohexanedione in 0.1 M sodium borate, pH 8.1.

k   Inactivation occurred at a rate of 10.9 M^{-1} min^{-1} under these conditions (compared to 4.0 M^{-1} min^{-1} with phenylglyoxal in bicarbonate/carbonate and 6.6 M^{-1} min^{-1} with 1,2-cyclohexanedione in 0.050 M borate, pH 9.0). Inactivation with 2,3-butanedione is not observed in 0.05 M bicarbonate/carbonate, pH 9.0 at 25°C; however there is reduced modification of arginine (0.4 residue per subunit as compared to 1.3 residues per subunit with 77% inactivation).

l   The majority of arginine modification could be reversed by the removal of reagent and borate solvent by dialysis vs. 0.05 M potassium phosphate, pH 7.8. Enzymic activity was also recovered as a result of the dialysis procedure. These investigators were able to obtain evidence supporting the selective modification of Arg[141] by either 2,3-butanedione, 1,2-cyclohexandione or phenylglyoxal.

m   The modification was performed at 22°C. These studies were performed with bacterial membrane preparations. Stoichiometry was not established. Analysis of the rates of inactivation suggested that inactivation was due to the modification of a single arginine residue. NADH, which stimulates the transhydrogenation of 3-acetylpyridine-NAD by NADPH, protects the enzyme from inactivation.

n   The modification was performed at 25°C. The extent of modification was determined by amino acid analysis after acid hydrolysis. The extent of modification reported was obtained after 75 min of reaction concomitant with 85% loss of activity. The presence of substrate, α-phosphoglycerate, reduced the extent of modification to 2 mol arginine per subunit with only 5% loss of catalytic activity.

o   A second-order rate constant of 0.0635 M^{-1} min^{-1} was obtained for the loss of enzymic activity upon reaction with 2,3-butanedione in 0.050 M borate, pH 7.0 at 25°C. This presumably reflects the modification of a single arginine residue (see Footnote p). The inactivation of the enzyme by 1,2-cyclohexanedione, methylglyoxal, and phenylglyoxal is compared with that by 2,3-butanedione (all at 10 mM in 0.05 M borate, pH 7.0). Butanedione is clearly most effective followed by phenylglyoxal, methylglyoxal, and 1,2-cyclohexanedione. The authors note that the enzyme under study, aldehyde reductase, can utilize methylglyoxal and phenylglyoxal as substrates, precluding their rigorous evaluation in this study.

p   Obtained by amino acid analysis after acid hydrolysis (6 N HCl, 110°C, 24 hr). The control preparation yielded a value of 17.8 ± 1 Arg while the modified enzyme yielded a value of 16.7 ± 1 Arg. The presence of cofactor yielded a preparation with 17.5 ± 1 Arg.

q   The reactions are reported at 25°C. Borate buffers could not be used since borate is a competitive inhibitor of the enzyme and prevents inactivation in bicarbonate buffer. Reaction with phenylglyoxal in the same solvent.

r   Reaction performed at 25°C. Stoichiometry established by amino acid analysis after acid hydrolysis (6 N HCl, 100°C, 20 hr). Arginine is the only amino acid modified under these reaction conditions. These values were obtained at 80% inactivation. The presence of ADP reduced activity loss to 55% with extent of arginine modification reduced to 0.4 to 0.5 residues.

s   Reaction performed at 25°C for 90 min. Stoichiometry determined by amino acid analysis after acid hydrolysis (6 N HCl, 110°C, 24 hr).

t   The use of isolated "membrane fraction" prevented the establishment of stoichiometry in these studies. Analysis of the dependence of reaction rate on concentration of 2,3-butanedione is consistent with the modification of a single arginine residue. As expected, the stability of modification is dependent upon the presence of borate. Gel filtration into HEPES (0.125 M, pH 7.0) and subsequent inactivation at 37°C resulted in the recovery of a substantial amount of catalytic activity. Similar results were obtained with imidazole and Tris buffers under similar reaction conditions. This reactivation does not occur when the incubation following gel filtration is performed at 0°C instead of 37°C.

u   A reaction rate with a second-order constant of k = 5.2 M^{-1} min^{-1} is obtained at 25°C. Inactivation is dependent on the presence of borate as inactivation is not observed with use of BICINE buffer. Dialysis vs. 0.025 M phosphate, pH 7.0 for 21 hr at 4°C results in an increase in activity of 14 to 85% while complete recovery of activity is achieved after 21 hr of dialysis.

Table 2 (continued)
USE OF 2,3-BUTANEDIONE TO MODIFY ARGINYL RESIDUES IN PROTEINS

[v]   Stoichiometry was not established for the reaction with 2,3-butanedione. As shown in Table 1, reaction with phenylglyoxal modifies approximately 4 of the 106 arginyl residues in each subunit of fatty acid synthetase. The loss of the biological activity as determined either by fatty acid synthetase activity, ketoreductase activity, or enoylreductase activity was considerably more rapid with phenylglyoxal than with 2,3-butanedione. It is noted that these reactions are performed in borate buffer for the studies with 2,3-butanedione and phosphate buffer for the studies with phenylglyoxal, both buffers at pH 7.6 with the reactions performed at 30°C.

[w]   Reactions were performed at 25°C. The modification of arginyl residues is associated with an approximate 70% loss of enzymatic activity. The presence of N-phenylpyridinium-2-aldoxine iodide reduces the extent of arginine modification by approximately 1 mol/mol of enzyme with concomitant protection of enzymatic activity. It should be noted that modification of this enzyme with phenylglyoxal results in the modification of 3 mol of arginine/mole enzyme with 17% loss of enzymatic activity (see Table 1). It is not clear as to when modification of a particular arginyl residue with the two reagents is a mutually exclusive event.

[x]   Reactions were performed at 25°C. Rigorous evaluation of the stoichiometry of the reaction is not available. Analysis of the dependence of first-order rate constants on reagent concentration (double-logarithmic relationships) is consistent with the modification of a single arginyl residue. The inactivation was reversed by 100-fold dilution into 0.05 M potassium phosphate, pH 8.5, at 25°C.

[y]   The inactivation of ornithine transcarbamylase is readily reversible in this solvent; the presence of borate precludes reactivation observed on dilution of modified enzyme in solvent. A value of 179 M^{-1} min^{-1} for the second-order rate constant for reaction of 2,3-butanedione with ornithine transcarbamylase under these conditions was recorded.

[z]   Obtained at 88% inactivation.

[aa] Reaction at 22°C.

[bb] Determined by amino acid analysis. The reaction is readily reversible, even in the presence of borate.

[cc] Determined by amino acid analysis. Constructed Scatchard plot shows that two residues were not available for modification with 2,3-butanedione.

[dd] Second-order rate constant, k = 2.95 M^{-1} min^{-1} in this solvent, assuming that loss in catalytic activity is a measure of reaction with arginine.

[ee] Stoichiometry was not established. Kinetic analysis suggests that inactivation of catalytic activity results from the modification of a single arginine residue.

[ff] Modification reaction was performed at 24°C. Very little inactivation is observed if the reaction is performed in Tris buffer at the same pH. Reactivation of enzyme modified in borate buffer is observed when the inactivated enzyme is diluted in borate buffer.

[gg] The reaction was performed at 25°C for 4 hr. The presence of the competitive inhibitor, 6-phosphogluconate, protected 1 mol of arginine/mol of enzyme from modification suggesting that there is a single arginine residue critical for catalytic activity. A 20-fold increase in inhibitor concentration resulted in the modification of greater than 95% of the total arginine residues.

[hh] Second-order rate constant of k = 7.5 M^{-1} min^{-1} at 25°C was obtained from the analysis of reaction rate data. pH Dependence study showed optimal rate of inactivation at pH 8.2.

[ii] Determined by amino acid analysis on 95 + % inactivated enzyme. Plotting loss of activity vs. arginine residues modified suggests that inactivation is due to the modification of a single arginine residue. Inactivation occurs with loss of sulfhydryl content.

[jj] Second-order rate constant of k = 13.57 M^{-1} min^{-1} obtained at 20°C. Inactivation much less rapid in 0.050 M HEPES, pH 8.3 ($T_{1/2}$ = 30 min in borate; 11.5 min in HEPES).

[kk] Reactions performed at 30°C. Modification associated with 80 to 90% inactivation. Reaction with phenylglyoxal (0.050 M sodium phosphate, 1 mM EDTA, pH 7.6) at 2000-fold molar excess led to the modification of 2 arginyl residues at a level of 90% inactivation. The extent of arginine was determined by spectrophotometric analysis (increase in absorbance at 250 nm, $\Delta\epsilon$ = 11,000 M^{-1} cm^{-1}; see Reference 33).

References for Table 2

1. **Riordan, J. F.**, Functional arginyl residues in carboxypeptidase A. Modification with butanedione, *Biochemistry*, 12, 3915, 1973.
2. **Fliss, H., Tozer, N. M., and Viswanatha, T.**, The reaction of chymotrypsin with 2,3-butanedione trimer, *Can. J. Biochem.*, 53, 275, 1975.
3. **Cipollo, K. L. and Dunlap, R. B.**, Essential arginyl residues in thymidylate synthetase, *Biochem. Biophys. Res. Commun.*, 81, 1139, 1978.
4. **McTigue, J. J. and Van Etten, R. L.**, An essential arginine residue in human prostatic acid phosphatase, *Biochim. Biophys. Acta*, 523, 422, 1978.
5. **Jordan, F. and Wu, A.**, Inactivation of purine nucleoside phosphorylase by modification of arginine residues, *Arch. Biochem. Biophys.*, 190, 699, 1978.

Table 2 (continued)

6. **Borders, C. L., Jr., Cipollo, K. L., and Jordasky, J. F.,** Role of arginyl residues in yeast hexokinase PII, *Biochemistry,* 17, 2654, 1978.

7. **Hayman, S. and Colman, R. F.,** Effect of arginine modification on the catalytic activity and allosteric activation by adenosine diphosphate of the diphosphopyridine nucleotide specific isocitrate dehydrogenase of pig heart, *Biochemistry,* 17, 4161, 1978.

8. **Enoch, H. G. and Strittmatter, P.,** Role of tyrosyl and arginyl residues in rat liver microsomal stearyl-coenzyme A desaturase, *Biochemistry,* 17, 4927, 1978.

9. **Malinowski, D. P. and Fridovich, I.,** Chemical modification of arginine at the active site of the bovine erythrocyte superoxide dismutase, *Biochemistry,* 18, 5909, 1979.

10. **Homyk, M. and Bragg, P. D.,** Steady-state kinetics and the inactivation by 2,3-butanedione of the energy-independent transhydrogenase of *Escherichia coli* cell membranes, *Biochim. Biophys. Acta,* 571, 201, 1979.

11. **Borders, C. L., Jr. and Zurcher, J. A.,** Rabbit muscle enolase also has essential argininyl residues, *FEBS Lett.,* 108, 415, 1979.

12. **Davidson, W. S. and Flynn, T. G.,** A functional arginine residue in NADPH-dependent aldehyde reductase from pig kidney, *J. Biol. Chem.,* 254, 3724, 1979.

13. **James, G. T.,** Essential arginine residues in human liver arylsulfatase A, *Arch. Biochem. Biophys.,* 197, 57, 1979.

14. **Grisham, C. M.,** Characterization of essential arginyl residues in sheep kidney $(Na^+ + K^+)$-ATPase, *Biochem. Biophys. Res. Commun.,* 88, 229, 1979.

15. **Pillai, R. P., Marshall, M., and Villafranca, J. J.,** Modification of an essential arginine of carbamate kinase, *Arch. Biochem. Biophys.,* 199, 16, 1980.

16. **Cipollo, K. L. and Dunlap, R. B.,** Essential arginyl residues in thymidylate synthetase from amethopterin-resistant *Lactobacillus casei, Biochemistry,* 18, 5537, 1979.

17. **Schrijen, J. J., Luyben, W. A. H. M., DePont, J. J. H. M., and Bonting, S. L.,** Studies on $(K^+ + H^+)$-ATPase. I. Essential arginine residue in its substrate binding center, *Biochim. Biophys. Acta,* 597, 331, 1980.

18. **Belfort, M., Maley, G. F., and Maley, F.,** A single functional arginyl residue involved in the catalysis promoted by *Lactobacillus casei* thymidylate synthetase, *Arch. Biochem. Biophys.,* 204, 340, 1980.

19. **Poulose, A. J. and Kolattukudy, P. E.,** Presence of one essential arginine that specifically binds the 2'-phosphate of NADPH on each of the ketoacyl reductase and enoyl reductase active sites of fatty acid synthetase, *Arch. Biochem. Biophys.,* 199, 457, 1980.

20. **Müllner, H. and Sund, H.,** Essential arginine residue in acetylcholinesterase from *Torpedo californica, FEBS Lett.,* 119, 283, 1980.

21. **Kuno, S., Toraya, T., and Fukui, S.,** Coenzyme B_{12}-dependent diol dehydrase: chemical modification with 2,3-butanedione and phenylglyoxal, *Arch. Biochem. Biophys.,* 205, 240, 1980.

22. **Marshall, M. and Cohen, P. P.,** Evidence for an exceptionally reactive arginyl residue at the binding site for carbamyl phosphate in bovine ornithine transcarbamylase, *J. Biol. Chem.,* 255, 7301, 1980.

23. **Dreyfus, M., Vandenbunder, B., and Buc, H.,** Mechanism of allosteric activation of glycogen phosphorylase probed by the reactivity of essential arginyl residues. Physicochemical and kinetic studies, *Biochemistry,* 19, 3634, 1980.

24. **Pande, J. and Myer, Y. P.,** The arginines of cytochrome c. The reduction-binding site for 2,3-butanedione and ascorbate, *J. Biol. Chem.,* 255, 11094, 1980.

25. **Tristram-Nagle, S. and Packer, L.,** Effects of arginine modification on the photocycle and proton pumping of bacteriorhodopsin, *Biochem. Int.,* 3, 621, 1981.

26. **Gomazkova, V. S., Stafeeva, and Severin, S. E.,** The role of arginine residues in the functioning of α-ketoglutarate dehydrogenase from pigeon breast muscle, *Biochem. Int.,* 2, 51, 1981.

27. **Wong, S. S. and Wong, L.-J.,** Evidence for an essential arginine residue at the active site of *Escherichia coli* acetate kinase, *Biochim. Biophys. Acta,* 660, 142, 1981.

28. **Chang, G.-G. and Huang, T.-M.,** Modification of essential arginine residues of pigeon liver malic enzyme, *Biochim. Biophys. Acta,* 660, 341, 1981.

29. **Lu, H. S., Talent, J. M., and Gracy, R. W.,** Chemical modification of critical catalytic residues of lysine, arginine and tryptophan in human glucose phosphate isomerase, *J. Biol. Chem.,* 256, 785, 1981.

30. **Fujioka, M. and Takata, Y.,** Role of arginine residue in saccharopine dehydrogenase (L-Lysine Forming) from baker's yeast, *Biochemistry,* 20, 468, 1981.

31. **Gacesa, P., Savitsky, M. J., Dodgson, K. S., and Olavesen, A. H.,** Modification of functional arginine residues in purified bovine testicular hyaluronidase with butane-2,3-dione, *Biochim. Biophys. Acta,* 661, 205, 1981.

32. **Boggaram, V. and Mannervik, B.,** Essential arginine residues in the pyridine nucleotide binding sites of glutathione reductase, *Biochim. Biophys. Acta,* 701, 119, 1982.

33. **Takahashi, K.,** Further studies on the reactions of phenylgloxal and related reagents with proteins, *J. Biochem.,* 81, 403, 1977.

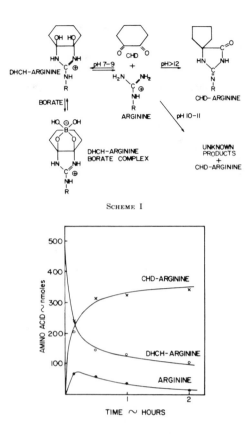

Scheme I

FIGURE 20. The reaction of arginine with 1,2-cyclohexanedione. Scheme I shows a representation of the reaction of 1,2-cyclohexanedione with arginine. The figure shows the conversion of DHCH-arginine to CHD-arginine in 0.5 *M* NaOH. Amino acids were determined on the amino acid analyzer. (From Patthy, L. and Smith, E. L., *J. Biol. Chem.*, 250, 557, 1975. With permission.)

nedione with arginine (Figure 20) forms N^5-(4-oxo-1,3-diazaspiro[4,4]non-2-yliodene)-L-ornithine (CHD-arginine), a reaction which cannot be reversed. Between pH 7.0 and pH 9.0 a compound is formed from arginine and 1,2-cyclohexanedione, N^7-N^8-(1,2-dihydroxycyclohex-1,2-ylene)-L-arginine (DHCH-arginine). This compound is stabilized by the presence of borate and is unstable in the presence of buffers such as Tris. This compound is readily converted back to free arginine in 0.5 *M* hydroxylamine, pH 7.0 (Figure 21). These authors have subsequently used this reagent to identify functional residues in bovine pancreatic ribonuclease A and egg white lysozyme.[39] Extent of modification of arginine residues in protein by 1,2-cyclohexanedione is generally assessed by amino acid analysis after acid hydrolysis. Under the conditions normally used for acid hydrolysis (6 *N* HCl, 110°C, 24 hr), the borate-stabilized reaction product between arginine and 1,2-cyclohexanedione is unstable and there is partial regeneration of arginine and the formation of unknown degradation products.[8] Acid hydrolysis in the presence of an excess of mercaptoacetic acid (20 μℓ/mℓ of hydrolysate) prevents the destruction of DHCD-arginine.[8] Table 3 lists some of the enzymes in which structure-function relationships have been studied by reaction with 1,2-cyclohexanedione and others are discussed below in comparison with phenylglyoxal and/or 2,3-butanedione. It is the purpose of the following section to discuss some selected studies on the use of the reagents discussed above for the modification of arginyl residues

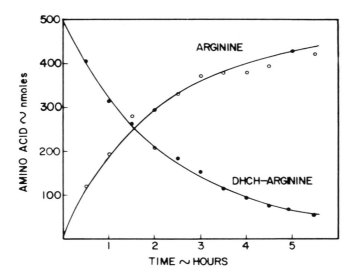

FIGURE 21.   Disappearance of DHCH-arginine and formation of arginine on treatment with neutral hydroxylamine. DHCH-arginine (0.1 *M*) was incubated in 0.5 *M* hydroxylamine, pH 7.0, at 37°C. The amino acids were determined on the amino acid analyzer. Both sets of data are first order with a half-time of 100 min. (From Patthy, L. and Smith, E. L., *J. Biol. Chem.*, 250, 557, 1975. With permission.)

Table 3
REACTION OF ARGINYL RESIDUES IN PROTEINS WITH 1,2-CYCLOHEXANEDIONE

Protein	Solvent	Reagent excess	Extent of modification	Ref.
Ribonuclease A	0.2 *M* sodium borate, pH 9.0	50,000	3/4	1
Lysozyme	0.2 *M* sodium borate, pH 9.0	50,000	11/11	1
Kunitz bovine trypsin inhibitor	0.2 *M* sodium borate, pH 9.0	—	5.5/6	2

References for Table 3

1. **Patthy, L. and Smith, E. L.,** Identification of functional arginine residues in ribonuclease A and lysozyme, *J. Biol. Chem.*, 250, 565, 1975.
2. **Menegatti, E., Ferroni, R., Benassi, C. A., and Rocchi, R.,** Arginine modification in Kunitz bovine trypsin inhibitor through 1,2-cyclohexanedione, *Int. J. Pept. Protein Res.*, 10, 146, 1977.

in proteins. We have chosen these particular studies *strictly on the basis* of the illustration of a *specific aspect* of the *chemistry* of a *given reagent(s)*. As will be apparent, a number of investigators use several reagents (e.g., phenylglyoxal and/or 2,3-butanedione) with a given protein.

Vallejos and co-workers[39] have examined the modification of a photophosphorylation factor in *Rhodospirillum rubrum* chromatophore with either 2,3-butanedione or phenylglyoxal as shown in Figure 22. The reactions were performed in 0.050 *M* borate, pH 7.8

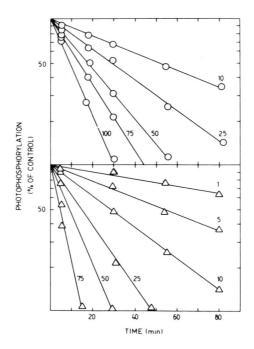

FIGURE 22.   Effect of 2,3-butanedione and phenylglyoxal on photophosphorylation in *Rhodospirillum rubrum* chromatophores. The experiments were performed in 50 mM borate buffer (pH 7.8) at 25°C with either 2,3-butanedione (o———o) or phenylglyoxal (△———△) at the concentrations indicated in the figure. The reactions were performed in the dark. (From Vallejos, R. H., Lescano, W. I. M., and Lucero, H. A., *Arch. Biochem. Biophys.*, 190, 578, 1978. With permission.)

(25°C). Stoichiometry is not reported but it is not unreasonable to suggest that the two reagents react at the same site, in which case phenylglyoxal is more effective. These reactions were performed in the dark. When the reaction with 2,3-butanedione is performed in the light there is an approximate 25-fold increase in the rate of inactivation. These investigators discuss this in terms of a conformational change in the chromatophore but do not consider possible photosensitization as described above. Homyk and Bragg[40] compared the effect of 2,3-butanedione and phenylglyoxal on the energy-independent transhydrogenase of *Escherichia coli*. The results of these experiments are shown in Figure 23. The reactions were performed in 0.050 M sodium borate, pH 7.8 at 22°C. Phenylglyoxal and 2,3-butanedione were of approximate equal effectiveness in reducing enzymatic activity. The insets show plots of the logarithm of the observed pseudo-first order rate constants vs. the logarithm of the inhibitor concentration. In this type of analysis a straight line should be obtained with a slope equal to the number of inhibitor molecules reacting with each active site to yield an inactive enzyme.[41,42] The analysis for phenylglyoxal yielded a slope of 1.1 while that for 2,3-butanedione gave a slope of 0.8. Therefore these experiments are consistent with the loss in catalytic activity resulting from the modification of one arginyl residue per active site of the enzyme. Also shown in Figure 23 is the protection by substrates and substrate analogues on the rate of inactivation by 2,3-butanedione.

The modification of hexokinase[43] by phenylglyoxal (Figure 24) is of interest since analysis of the dependence of reaction rate on reagent concentration suggests the formation of a protein-phenylglyoxal complex prior to the modification of arginine. Note also that a stoichiometry of 1:1 is suggested based on [^{14}C] phenylglyoxal incorporation while the original

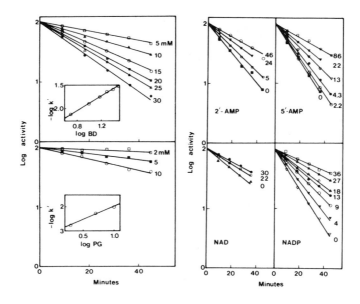

FIGURE 23.   Comparison of the modification of an energy-dependent transhydrogenase from *Escherichia coli* cell membranes with 2,3-butanedione and phenylglyoxal. On the left is shown the kinetics of inactivation of the energy-dependent transhydrogenase by 2,3-butanedione (BD) (upper panel) and phenylglyoxal (PG) (lower panel). The membranes were prepared in 50 mM sodium borate, pH 7.8, and incubated at 23°C at a concentration of 1.1 mg and 1.96 mg protein/mℓ with the indicated concentration of 2,3-butanedione and phenylglyoxal, respectively. Samples were withdrawn and assayed at timed intervals. Activity is expressed as a percentage of the control activity taken at the onset of incubation. The insets show the relationship between the pseudo first-order rate constant of inactivation (k^1, expressed as min^{-1}) and the inhibitor concentration (expressed as mM). The effect of substrates and substrate analogs on the inactivation of the energy-dependent transhydrogenase by 2,3-butanedione is shown on the right. Membranes at a concentration of 1.0 (top panels) and 1.4 (bottom panels) mg protein/mℓ in 50 mM sodium borate, pH 7.8, were incubated at 22°C with 53.7 mM 2,3-butanedione, in the absence or presence of the indicated millimolar concentrations of substrates and substrate analogs. Samples were withdrawn at timed intervals for assay. Activity is expressed as a percentage of the control activity taken at the onset of incubation. (From Homyk, M. and Bragg, P. D., *Biochim. Biophys. Acta,* 571, 201, 1979. With permission.)

studies with free arginine[2] suggested 2:1 stoichiometry. The reaction of phenylglyoxal with arginyl residues in a myosin fragment[44] (Figures 25 and 26) also suggests ''saturation kinetics'' with the formation of a reagent-protein complex prior to modification of an arginine residue. The effect of bicarbonate on the reaction of phenylglyoxal with arginine has been discussed above. Fonda and co-workers[45] have extended these observations to a consideration of the modification of arginyl residues in glutamate decarboxylase. The holoenzyme is resistant to inactivation by a variety of reagents specific for arginine while the apoenzyme is susceptible. Phenylglyoxal was the most effective (k$_2$ = 107.68 M^{-1} min^{-1} in 0.125 M bicarbonate, pH 7.5) of the reagents tested. Figure 27 shows the time course of inactivation as a function of pH. The comparison of different buffers for this reaction is shown in Figure 28 while the effect of bicarbonate concentration is shown in Figure 29. As discussed by these investigators, the bicarbonate effect may be a *general* effect but it is our experience that the modification of arginyl residues in proteins proceeds more rapidly in this buffer system.

The studies of Davidson and Flynn[46] on the modification of an arginyl residue in an

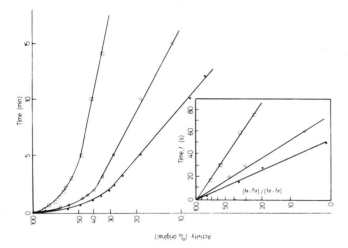

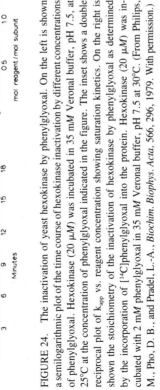

FIGURE 25. Rates of inactivation of K^+ EDTA-ATPase of myosin subfragment 1 as a function of phenylglyoxal concentration. Myosin subfragment 1 (obtained by the limited proteolysis of rabbit skeletal muscle myosin filaments by chymotrypsin) (1 mg/mℓ) was allowed to react with 1 mM ($\square$), 2 mM ($\triangle$), or 3 mM ($\bullet$) phenylglyoxal in 0.1 M potassium bicarbonate, pH 8, at 25°C. ATPase tests were performed at pH 7.5 on 50-$\mu\ell$ portions at the indicated times. The biphasic reaction is described in terms of K_1, the faster rate constant for partial inactivation and K_2, the slower rate constant for total inactivation. Inset: determination of the pseudo first-order rate constants K_1 for the rapid phase of inactivation. V_o = original activity, V_t = activity at time t; for each phenylglyoxal concentration, V_F is the velocity at the end of the faster phase of the reaction. (From Mornet, D., Pantel, P., Audemard, E., and Kassab, R., *Eur. J. Biochem.*, 100, 421, 1979. With permission.)

FIGURE 24. The inactivation of yeast hexokinase by phenylglyoxal. On the left is shown a semilogarithmic plot of the time course of hexokinase inactivation by different concentrations of phenylglyoxal. Hexokinase (20 μM) was incubated in 35 mM Veronal buffer, pH 7.5, at 25°C at the concentration of phenylglyoxal indicated in the figure. The inset shows a double reciprocal plot of k_{app} vs. reagent concentration showing saturation kinetics. On the right is shown the stoichiometry of the inactivation of hexokinase by phenylglyoxal as determined by the incorporation of [^{14}C]phenylglyoxal into the protein. Hexokinase (20 μM) was incubated with 2 mM phenylglyoxal in 35 mM Veronal buffer, pH 7.5 at 30°C. (From Philips, M., Pho, D. B., and Pradel, L.-A., *Biochim. Biophys. Acta*, 566, 296, 1979. With permission.)

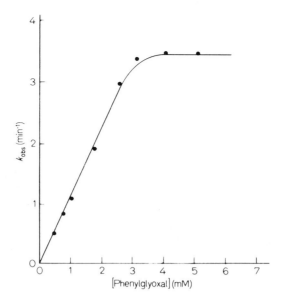

FIGURE 26. Dependence of the pseudo first-order rate constants for K$^+$ EDTA-ATPase inactivation on the concentration of phenylglyoxal. The rate constants were calculated from the inset in Figure 28. (From Mornet, D., Pantel, P., Audemard, E., and Kassab, R., *Eur. J. Biochem.*, 100, 421, 1979. With permission.)

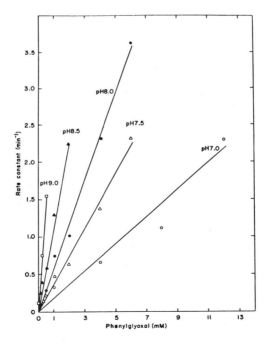

FIGURE 27. The effect of pH on the apparent first-order rate constants of glutamate apodecarboxylase inactivation by phenylglyoxal. Apoenzyme (33 μM) was incubated with varying concentrations of phenylglyoxal in 83 mM bicarbonate buffer. The pH of each solution was determined after the addition of apoenzyme in 10 mM pyridine-Cl, pH 4.6. (From Cheung, S.-T. and Fonda, M. L., *Arch. Biochem. Biophys.*, 198, 541, 1979. With permission.)

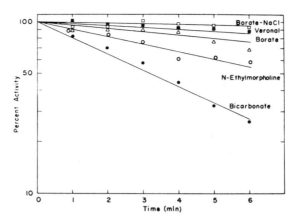

FIGURE 28. The effect of buffer on the inactivation of glutamate apodecarboxylase by phenylglyoxal at pH 7.5. The apoenzyme was incubated with 2 mM phenylglyoxal in 35 mM borate buffer (with or without 1 M NaCl), 50 mM Veronal buffer, 140 mM N-ethylmorpholine-Cl buffer, or 83 mM sodium bicarbonate buffer. (From Cheung, S.-T. and Fonda, M. L., *Arch. Biochem. Biophys.*, 199, 457, 1980. With permission.)

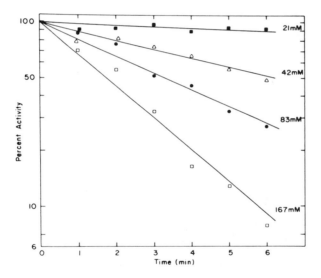

FIGURE 29. The effect of bicarbonate concentration on the inactivation of glutamate apodecarboxylase by 2 mM phenylglyoxal at pH 7.5. The final concentrations of bicarbonate are indicated. (From Cheung, S.-T. and Fonda, M. L., *Arch. Biochem. Biophys.*, 198, 541, 1979. With permission.)

aldehyde reductase provided another evaluation of several different reagents. Figure 30 shows the results of an experiment performed in 0.050 M sodium borate, pH 7.0, at 25°C in which 1,2-cyclohexanedione, methylglyoxal, phenylglyoxal, and 2,3-butanedione are compared. 2,3-Butanedione is the most potent inactivator with phenylglyoxal being somewhat less effective while 1,2-cyclohexanedione is least effective. Another comparison of phenylglyoxal and 2,3-butanedione is provided from the studies of Poulose and Kolattukudy[47] on the participation of arginyl residues in fatty acid synthetase. Figure 31 contains a comparison

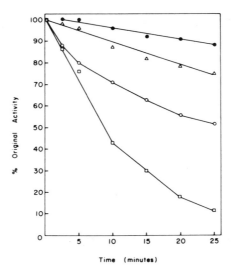

FIGURE 30.   Modification of an aldehyde reductase by several reagents specific for reaction with arginine. Shown is the inactivation of pig kidney aldehyde reductase by 10 mM 1,2-cyclohexanedione (●), 10 mM methylglyoxal (△), 10 mM phenylglyoxal (○), and 10 mM 2,3-butanedione (□). The enzyme (0.1 mg/mℓ) was incubated with the various reagents in 50 mM borate buffer, pH 7.0, at 25°C. Portions were removed at the indicated times, diluted into ice cold 0.1 M sodium phosphate, pH 7.0, and assayed for catalytic activity. (From Davidson, W. S. and Flynn, T. G., *J. Biol. Chem.*, 254, 3724, 1979. With permission.)

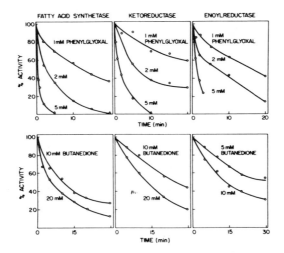

FIGURE 31.   Time course of inactivation of ketoacyl reductase, enoyl reductase, and the overall activity of fatty acid synthetase by phenylglyoxal and 2,3-butanedione. Incubation with phenylglyoxal was carried out at a protein concentration of 4.1, 4.3, and 3.3 mg/mℓ for fatty acid synthetase, ketoacyl reductase, and enoyl reductase, respectively; incubations with 2,3-butanedione were done at 0.33, 0.165, and 1.63 mg/mℓ, respectively. The modification with phenylglyoxal was performed in 100 mM sodium phosphate, pH 7.6, containing 0.5 mM dithiothreitol and 1.0 mM EDTA at 30°C. The modification with 2,3-butanedione was performed in 20 mM borate containing 200 mM KCl, 1.0 mM dithiothreitol, and 1.0 mM EDTA at pH 7.6 at 30. (From Poulose, A. J. and Kolattukudy, P. E., *Arch. Biochem. Biophys.*, 199, 457, 1980. With permission.)

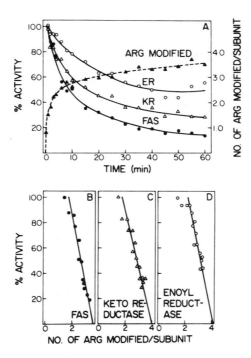

FIGURE 32. Analysis of the modification of arginine residues in fatty acid synthetase by phenylglyoxal. Shown in panel A is the time course of inactivation of the three enzymatic activities and the number of arginine residues modified with 1 mM [2-^{14}C] phenylglyoxal in 100 mM sodium phosphate buffer (pH 7.0) containing 0.5 mM dithiothreitol and 1.0 mM EDTA. The reactions were performed at a protein concentration of 8.8 mg/mℓ at 30°C. Panels B to D show the stoichiometry of modification for overall fatty acid synthetase (FAS), keto reductase, and enoyl reductase. (From Poulose, A. J. and Kolattukudy, P. E., *Arch. Biochem. Biophys.*, 199, 457, 1980. With permission.)

of the effect of these two reagents on the various catalytic activities of this multifunctional enzyme. These experiments with phenylglyoxal were performed in 0.1 M sodium phosphate, pH 7.6, containing 1.0 M EDTA and 0.5 mM dithioerythritol, while those with 2,3-butanedione were performed in 0.020 M borate, 0.200 M KCl, pH 7.6, containing 1.0 mM EDTA and 1.0 mM dithioerythritol. Phenylglyoxal was more potent than 2,3-butanedione in the inactivation of the three catalytic activities. Figure 32 provides a further analysis of the reaction of fatty acid synthetase with phenylglyoxal. It is noted that only four arginyl residues out of a total of 106 arginyl residues per subunit are modified under these reaction conditions.

The investigation of the reaction of *Lactobacillus casei* thymidylate synthetase with phenylglyoxal is an interesting example of the use of this reagent.[58] Figure 33 describes the pH dependence (*N*-ethylmorpholine buffers) for the inactivation. A time-course study for the modification of arginyl residues by phenylglyoxal (350–fold molar excess, 0.200 M N-ethylmorpholine, pH 8.2, 25°C) in the presence and absence of a competitive inhibitor (2'-deoxyuridylate) is shown in Figure 34. The reactivation of the modified enzyme in the presence and absence of hydroxylamine is shown in Figure 35.

Studies from Cooperman's laboratory[48] on the modification of yeast inorganic pyrophosphatase presented some interesting data on the stability of the reaction product between arginine and phenylglyoxal. Figure 36 shows the change in the UV spectrum of the adduct between 2 molecules of phenylglyoxal and 1 molecule of arginine on incubation in 0.1 M

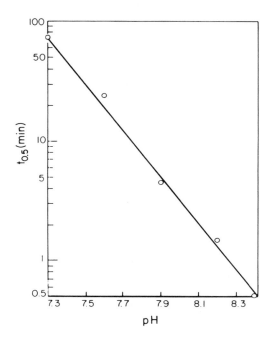

FIGURE 33.   pH Dependence of inactivation of thymidylate synthetase by phenylglyoxal.
Enzyme (7.1 μM) was incubated under argon at 25°C in 200 mM N-ethylmorpholine at the
pH indicated in the presence of 5 mM phenylglyoxal. The $t_{0.5}$ at each of the indicated pH
values was determined from a series of semilog plots of inactivation vs. time. (From Belfort,
M., Maley, G. F., and Maley, F., *Arch. Biochem. Biophys.*, 204, 340, 1980. With permission.)

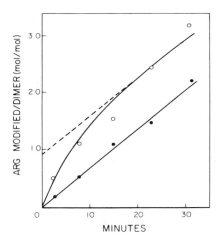

FIGURE 34.   The rate of modification of arginine in thymidylate synthetase by phenylglyoxal. Enzyme (14.3 μM) in 200 mM N-ethylmaleimide (pH 8.2) was preincubated for 30
min at 25°C in the absence (○) and presence (●) of a 200-fold molar excess of deoxyuridylate
(2.86 mM). At 0 min, phenylglyoxal was added to a concentration of 5 mM (350-fold molar
excess over enzyme). At the times indicated, 100 $\mu\ell$ portions were added to 1 mℓ of cold
0.1 N HCl in acetone to halt arginine modification and to precipitate the enzyme. The number
of arginine residues modified was calculated by comparison with an unmodified control.
(From Belfort, M., Maley, G. F., and Maley, F., *Arch. Biochem. Biophys.*, 204, 340, 1980.
With permission.)

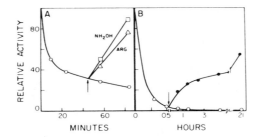

FIGURE 35. Reactivation of phenylglyoxal-treated thymidylate synthetase. A) The enzyme (14 μ*M*) was inactivated with a 30-fold excess of phenylglyoxal in 200 m*M* *N*-ethylmorpholine (pH 8.2) at 25°C. After 45 min, hydroxylamine (□) or arginine (△) were added to portions of native and phenylglyoxal-treated enzyme to a final concentration of 0.33 and 0.1 *M*, respectively. Activity is expressd in each case relative to a corresponding control containing no phenylglyoxal. (B) Enzyme (2.8 μ*M*) was incubated in the presence of a 140-fold molar excess of phenylglyoxal in 200 m*M* *N*-ethylmorpholine (pH 8.2) at 25°C. After 30 min a portion of the modified enzyme was passed through a column of Sephadex G-25 (arrow) equilibrated with 200 m*M* *N*-ethylmorpholine buffer to free the enzyme of unbound phenylglyoxal. The modified enzyme solutions before (○) and after gel filtration (●) were incubated at 25°C in an argon atmosphere and periodically assayed to determine the extent of reactivation. Activity is expressed relative to that of unmodified enzyme which was treated similarly. (From Belfort, M., Maley, G. F., and Maley, F., *Arch. Biochem. Biophys.*, 204, 340, 1980. With permission.)

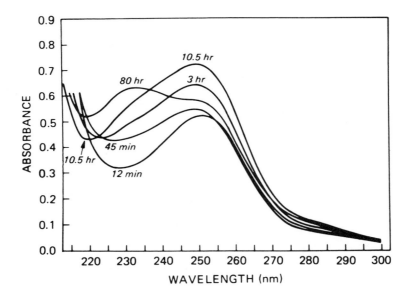

FIGURE 36. Ultraviolet absorption spectral changes on hydrolysis of the reaction product of phenylglyoxal and arginine. Shown are the ultraviolet spectral changes on hydrolysis of N°–acetyl-L-(diphenylglyoxal)arginine (NAcArg(PhGx)$_2$) in 0.1 *M* sodium phosphate buffer, pH 8.0, at 25°C. Spectra were obtained at the times indicated in the figure. (From Bond, M. W., Chiu, N. Y., and Cooperman, B. S., *Biochemistry*, 19, 94, 1980. With permission.)

sodium phosphate, pH 8.0. These data are consistent with a model for reaction where there is a rapid dissociation to form free arginine and reagent followed by the formation of a new reaction product with undefined stoichiometry. This study also provided an excellent example of the rigorous evaluation of reaction stoichiometry as shown in Figure 37.

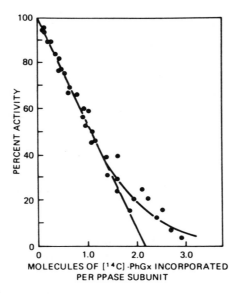

FIGURE 37.   Stoichiometry for the modification of yeast inorganic pyrophosphatase by phenylglyoxal. Shown is the correlation of [^{14}C]phenylglyoxal (PhGx) incorporation with pyrophosphatase inactivation. Experimental conditions: 0.08 M N-ethylmorpholine acetate (pH 7.0), 0.037 M PhGx, PPase (4 mg/mℓ), 35°C. (From Bond, M. W., Chiu, N. Y., and Cooperman, B. S., *Biochemistry,* 19, 94, 1980. With permission.)

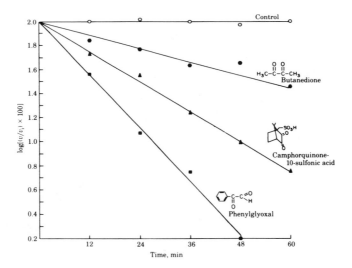

FIGURE 38.   The inactivation of choline acetyltransferase (ChoAcTase) by several reagents specific for the modification of arginine. The reagents indicated in the figure were present at a concentration of 10 mM in 50 mM HEPES, pH 7.8, at 25°C; v_i and v, initial enzyme activity and enzyme activity at any time point. (From Mautner, H. G., Pakula, A. A., and Merrill, R. E., *Proc. Natl. Acad. Sci. U.S.A.,* 78, 7449, 1981. With permission.)

Studies[49] on the modification of arginyl residues in choline acetyltransferase provide a further comparison of 2,3-butanedione and phenylglyoxal as well as a study of camphorquinone-10-sulfonic acid. Figure 38 compares the rate of inactivation of choline acetyltrans-

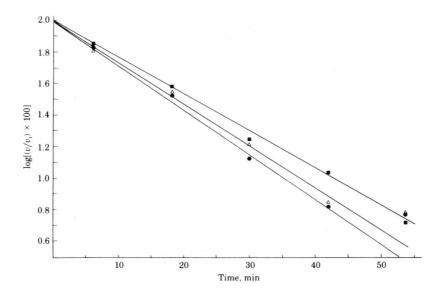

FIGURE 39.   Inactivation of choline acetyltransferase by camphorquinone-10-sulfonic acid as a function of pH. The reactions were performed at pH 7.0 (●), pH 8.0 (△), and pH 9.0 (■) in borate buffers. (From Mautner, H. G., Pakula, A. A., and Merrill, R. E., *Proc. Natl. Acad. Sci. U.S.A.*, 78, 7449, 1981. With permission.)

ferase by the three reagents. Phenylglyoxal is the most effective inactivator but it must be noted that the studies were performed in 0.040 M HEPES, pH 7.8 (25°C) in the *absence* of borate which is critical[36] for reaction of arginyl residues with 2,3-butanedione. The pH dependence for the modification of choline acetyltransferase by camphorquinone-10-sulfonic acid is shown in Figure 39.

Hayman and Colman[50] have studied the modification of arginyl residues in isocitrate dehydrogenase. Figure 40 shows the dependence of first-order rate constant for inactivation of catalytic activity as a function of 2,3-butanedione concentration (0.050 M MES, pH 6.2 containing 20% glycerol and 2.1 mM MnSO$_4$, 30°C). A second-order rate constant of 0.21 min^{-1} M^{-1} was obtained. The stoichiometry of the reaction is shown in Figure 41. The study[19] from Strittmatter's laboratory on the modification of arginine in rat liver microsomal stearylcoenzyme A desaturase provides an excellent example of the effect of borate on the 2,3-butanedione modification. Figure 42 shows the reversibility of enzyme inactivation by 2,3-butanedione on transfer from borate into Tris buffer and Figure 43 describes the correlation between inactivation, reactivation, and the extent of arginine modification. Another example of the effect of borate buffer on the reaction of 2,3-butanedione with protein is provided by the work of Varimo and Londesborough.[51] Figure 44 shows the pH dependence for the inactivation of yeast adenylate cyclase in borate buffer and HEPES buffer. Phenylborate or *m*-aminophenylborate buffers are also effective with 2,3-butanedione as demonstrated by studies on the modification of arginyl residues in bovine erythrocyte superoxide dismutase[52] as shown in Figure 45. In general the modification of arginyl residues by 2,3-butanedione proceeds more effectively at alkaline pH as illustrated by results with an ATPase[53] (Figure 46) and saccharopine dehydrogenase[54] (Figure 47).

Studies by Aurebekk and Little[55] compared the rate of inactivation of phospholipase c (*Bacillus cereus*) by 2,3-butanedione, 1,2-cyclohexanedione and phenylglyoxal in 0.020 M sodium borate, pH 7.0, as shown in Figure 48. The rate of reaction with 1,2-cyclohexanedione was intermediate between the other two reagents and was examined in further detail.

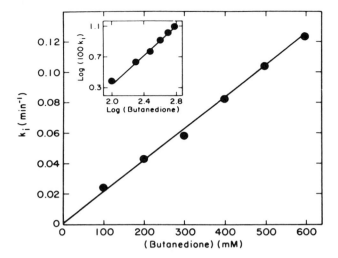

FIGURE 40. The loss of catalytic activity in porcine diphosphopyr-idine nucleotide-specific isocitrate dehydrogenase on modification with 2,3-butanedione. Shown is the dependence of k_i (pseudo first-order rate constant) on reagent concentration. The reactions were 50 mM in MES (pH 6.2), 20% in glycerol, and contained 2.1 mM MnSO$_4$ at 30°C. The 2,3-butanedione concentration is indicated in the figure. The rate constant for each reagent concentration was determined from a semilogarithmic plot of the loss of catalytic activity as a function of time. The second-order rate constant is 0.21 M^{-1} min^{-1}. The inset shows a plot of log k_i vs. log [2,3-butanedione], with a slope of 0.95. (From Hayman, S. and Colman, R. F., *Biochemistry*, 17, 4161, 1978. With permission.)

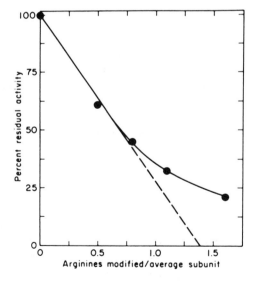

FIGURE 41. Stoichiometry for the modification of DPN-dependent isocitrate dehydrogen-ase with 2,3-butanedione. The enzyme was incubated for various times up to 73 min. at 30°C in 50 mM MES (pH 6.2) containing 2.1 mM MnSO$_4$ and 0.1 M 2,3-butanedione. Portions of the reacted enzyme containing 0.2 mg protein were diluted with an equal volume of 2 N HCl at 0°C to both stop the reaction and prevent the regeneration of free arginine. The modified enzyme was then dialyzed overnight at 4°C against 1 N HCl. The samples were dried *in vacuo* over solid NaOH and subjected to acid hydrolysis. Unmodified enzyme had an arginine content of 13.4 residues per average subunit of 40,000 molecular weight. (From Hayman, S. and Colman, R. F., *Biochemistry*, 17, 4161, 1978. With permission.)

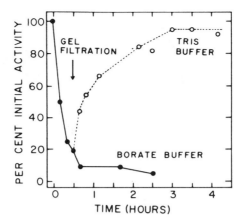

FIGURE 42. The reversal of 2,3-butanedione inactivation of stearylcoenzyme A desaturase. Shown is the reversibility of desaturase inactivation by removal of borate. Desaturase (16 μM, specific activity = 190 units/mg) was treated with 2,3-butanedione (40 mM in 50 mM sodium borate (pH 8.1); after 30 min at 25°C, 1 mℓ of this mixture was filtered through Sephadex G-25 (20 × 1 cm) equilibrated either with the same borate buffer or 20 mM Tris-acetate/100 mM NaCl (pH 8.1). The enzyme, which eluted in the void volume within 3 to 4 min, was incubated at 25°C, and portions were withdrawn at the times indicated for measurement of desaturase activity: (●) borate buffer; (○) Tris buffer. (From Enoch, H. G. and Strittmatter, P., *Biochemistry,* 17, 4927, 1978. With permission.)

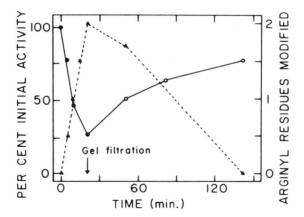

FIGURE 43. The correlation of the reversible modification of stearylcoenzyme A desaturase activity with the modification of arginyl residues. The treatment of desaturase with 2,3-butanedione and the removal of reagent and borate by gel filtration were carried out as described under Figure 42. At various times, samples of the reaction mixture were chilled rapidly to 0°C and used immediately for the measurement of desaturase activity (●) before and (○) after gel filtration, and arginine, (▲) before and (△) after gel filtration. (From Enoch, H. G. and Strittmatter, P., *Biochemistry,* 17, 4927, 1978. With permission.)

The pseudo first-order rate constant for the inactivation (in 0.2 *M* sodium borate, pH 8.0, 37°C) was plotted vs. 1,2-cyclohexanedione concentration (Figure 49); a straight line through the origin was obtained. A second-order rate constant ($0.42\ M^{-1}\ \text{min}^{-1}$) was obtained for these reaction conditions. The stoichiometry for this reaction is shown in Figure 50. Although neutral hydroxylamine will reverse the 1,2-cyclohexanedione-arginine adduct,[8] it was not

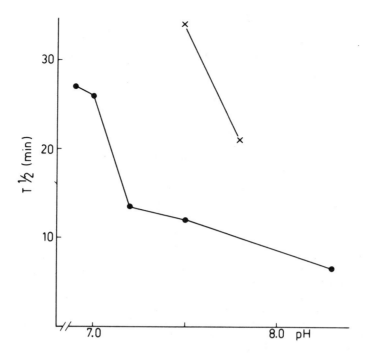

FIGURE 44.   pH Dependence for the inactivation of adenyl cyclase by 2,3-butanedione in two different buffers. Shown are the half-inactivation times for adenylate cyclase at 30°C with 20 mM 2,3-butanedione in 25 mM borate (●) and 25 mM HEPES (x) as a function of pH. Protein concentration during the incubation was 30 mg/mℓ. (From Varimo, K. and Londesborough, S., *FEBS Lett.*, 106, 153, 1979. With permission.)

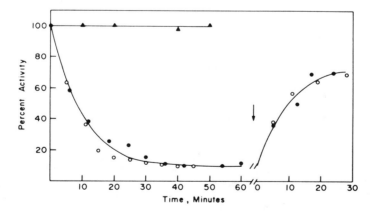

FIGURE 45.   Inactivation of bovine erythrocyte superoxide dismutase by 2,3-butanedione in phenylborate and *m*-aminophenylborate buffer. Native bovine erythrocyte superoxide dismutase (2.1 μM) was incubated in 1.0 mℓ of the following solutions at 25°C: 50 mM phenylborate, pH 9.0 (▲); 11.3 mM 2,3-butanedione, and 50 mM phenylborate, pH 9.0 (●); 11.3 mM 2,3-butanedione and 25 mM *m*-aminophenylborate, pH 9.0 (○). At the arrow, portions of the inactivated samples were diluted into 3.0 mℓ assay buffer containing 0.1 M mannitol (the mannitol was added to scavenge the borate) and incubated at 25°C prior to the addition of xanthine oxidase and determination of residual activity. (From Malinowski, D. P. and Fridovitch, I., *Biochemistry*, 18, 5909, 1979. With permission.)

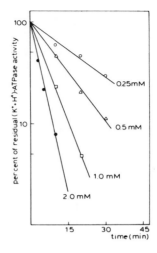

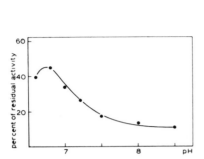

FIGURE 46. The inactivation of $(K^+ + H^+)$ ATPase by 2,3-butanedione. On the left is shown the inactivation by 2,3-butanedione as a function of pH. $(K^+ + H^+)$-ATPase preparation (0.5 mg/mℓ) was incubated for 20 min at 37°C with 0.5 mM 2,3-butanedione in 125 mM borate buffer containing 5 mM MgCl$_2$, previously brought to the indicated pH values with 5 M NaOH. $(K^+ + H^+)$-ATPase activity is expressed as percent of a control preparation without 2,3-butanedione. On the right is shown the inactivation by 2,3-butanedione as a function of time. $(K^+ H^+)$-ATPase preparation (0.5 mg protein/mℓ) was incubated at 37°C with the indicated concentrations of 2,3-butanedione in 125 mM sodium borate (pH 7.0), 5 mM MgCl$_2$. Enzyme activity is expressed as percent of control activity without butanedione. (From Schrijen, J. J., Luyben, W. A. H. M., DePont, J. J. H. M., and Bonting, S. L., *Biochim. Biophys. Acta,* 597, 331, 1980. With permission.)

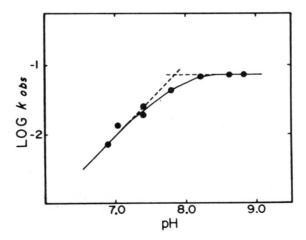

FIGURE 47. The effect of pH on the apparent first-order rate constant for the inactivation of saccharopine dehydrogenase by 2,3-butanedione. The enzyme (0.7 nmol) was incubated with 11.4 mM 2,3-butanedione in 0.1 mℓ of 0.08 M HEPES buffer at the pH values indicated. The buffers contained 0.2 M KCl and 10 mM borate in addition. Values of the apparent first-order rate constants (k_{obsd}) were obtained from the pseudo first-order kinetic plots. (From Fujioka, M. and Takata, Y., *Biochemistry,* 20, 468, 1981. With permission.)

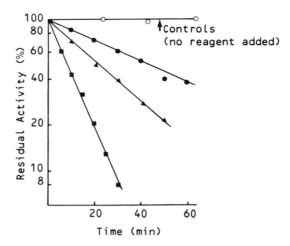

FIGURE 48.   The inactivation of phospholipase c by arginine reagents. Enzyme (0.25 mg/mℓ) incubated with phenylglyoxal (32 mM) in 0.02 M sodium borate buffer (pH 7.0) at 22°C (●). Enzyme (0.22 mg/mℓ) incubated with 2,3-butanedione (50 mM) in 0.06 M sodium borate buffer (pH 7.5) at 22°C (■). Enzyme incubated with 1,2-cyclohexanedione (85 mM) in 0.2 M sodium borate buffer (pH 8.0) at 37°C (▲). (From Aurebekk, B. and Little, C., *Int. J. Biochem.*, 8, 757, 1977. With permission.)

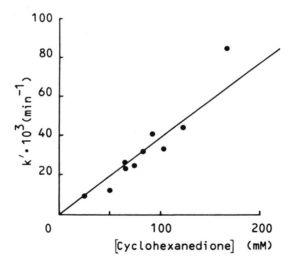

FIGURE 49.   The inactivation of phospholipase c by 1,2-cyclohexanedione. Shown is the effect of altering the 1,2-cyclohexanedione concentration on the pseudo first-order rate constant (k'). The enzyme (0.32—0.37 mg/mℓ) was incubated in 0.2 M sodium borate, pH 8.0, at 37°C with the indicated concentration of 1,2-cyclohexanedione and the pseudo first-order rate constants for the inactivation measured. (From Aurebekk, B. and Little, C., *Int. J. Biochem.*, 8, 757, 1977. With permission.)

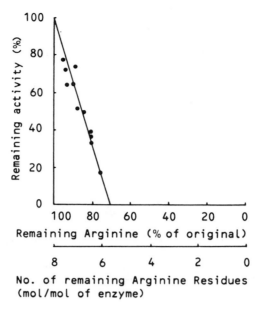

FIGURE 50. Stoichiometry for the modification of phospholipase c by 1,2-cyclohexanedione. Shown is the relationship between enzyme inactivation and the modification of arginine residues. Enzyme was inactivated to different extents by using varying concentrations of 1,2-cyclohexanedione and different reaction times in 0.2 *M* sodium borate, pH 8.0, at 37°C. After dialysis (twice) against 0.2 *M* sodium borate buffer (pH 8.0), the arginine content of the enzyme was measured by amino acid analysis after acid hydrolysis (6 *N* HCl) in the presence of mercaptoethanol. (From Aurebekk, B. and Little, C., *Int. J. Biochem.*, 8, 757, 1977. With permission.)

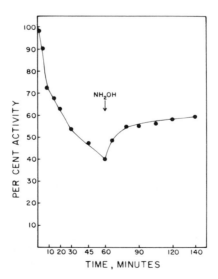

FIGURE 51. Reversibility of the inactivation of an acid phosphatase by 1,2-cyclohexanedione. Acid phosphatase in 50 m*M* borate buffer (pH 8.1) was modified at 30°C by 50 m*M* 1,2-cyclohexanedione. At the time indicated by the arrow, neutral hydroxylamine solution was added to a final concentration of 0.2 *M*. (From McTigue, J. J. and Van Etten, R. L., *Biochim. Biophys. Acta,* 523, 422, 1978. With permission.)

possible to use this observation in these studies on phospholipase c since the activity was lost in the presence of hydroxylamine. Only partial reactivation with neutral hydroxylamine is observed with 1,2-cyclohexanedione-modified prostatic acid phosphatase[56] (Figure 51).

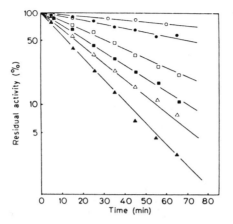

FIGURE 52.   Rate of inactivation of D-amino acid oxidase by 1,2-cyclohexanedione. 500 μg (12.6 nmol) of enzyme was incubated at 30°C in a final volume of 0.5 mℓ containing 200 m*M* borate buffer, pH 8.8, and various concentrations of 1,2-cyclohexanedione. Portions were removed at the indicated times and assayed for catalytic activity. The molar ratios of 1,2-cyclohexanedione to enzyme were 200:1 (●); 500:1 (□); 1000:1 (■); 1500:1 (△), and 2500:1 (▲). The curve (○) shows the loss of activity when the 1,2-cyclohexanedione–treated enzyme (inhibitor:enzyme molar ratio, 2500:1) was preincubated with 6 μmole of benzoate (a competitive inhibitor) for 5 min. at 30°C. (From Ferti, C., Curti, B., Simonetta, M. P., Ronchi, S., Galliano, M., and Minchiotti, L., *Eur. J. Biochem.*, 119, 553, 1981. With permission.)

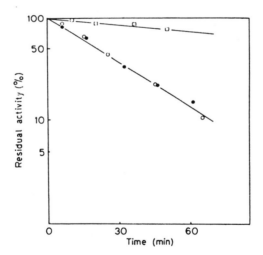

FIGURE 53.   Comparison of the rates of inactivation of D-amino acid oxidase holoenzyme and apoprotein in borate buffer and of holoenzyme in Tris buffer. (○), Holoenzyme in borate buffer with a ratio of 1,2-cyclohexanedione to enzyme of 1000:1; (●), apoprotein in borate buffer with the ratio of 1,2-cyclohexanedione to enzyme of 1000:1; and (□), holoenzyme in 100 m*M* Tris, pH 8.8 with a ratio of 1,2-cyclohexanedione to enzyme of 1000:1. Conditions other than indicated as listed under Figure 52. (From Ferti, C., Curti, B., Simonetta, M. P., Ronchi, S., Galliano, M., and Minchiotti, L., *Eur. J. Biochem.*, 119, 553, 1981. With permission.)

Studies of the modification of D-amino acid oxidase with 1,2-cyclohexanedione[57] (Figures 52 and 53) are of interest since a plot of the observed pseudo first-order rate constant vs. reagent concentration (Figure 54) is consistent with the formation of an enzyme-inhibitor complex formation prior to adduct formation.

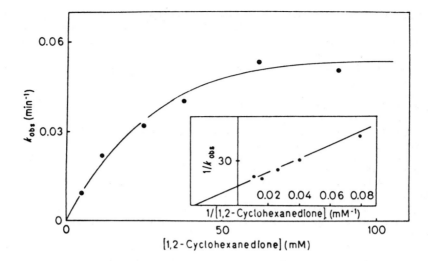

FIGURE 54. The effect of reagent concentration on the inactivation of D-amino acid oxidase by 1,2-cyclohexanedione. Shown is the effect of 1,2-cyclohexanedione concentration on the rate of D-amino acid oxidase inactivation in 200 mM borate buffer, pH 8.8, at 30°C. The values for k_{obs} (psuedo first-order rate constant) were determined from slopes of semilogarithmic plots similar to those shown in Figure 52 and plotted against the concentration of 1,2-cyclohexanedione. The inset shows a double-reciprocal plot of the data shown in the main figure. Note the suggestion of saturation kinetics in this figure. (From Ferti, C., Curti, B., Simonetta, M. P., Ronchi, S., Galliano, M., and Minchiotti, L., *Eur. J. Biochem.*, 119, 553, 1981. With permission.)

REFERENCES

1. **Yankeelov, J. A., Jr.,** Modification of arginine by diketones, *Meth. Enzymol.*, 25, 566, 1972.
2. **Takahashi, K.,** The reaction of phenylglyoxal with arginine residues in proteins, *J. Biol. Chem.*, 243, 6171, 1968.
3. **Yankeelov, J. A., Jr., Mitchell, C. D., and Crawford, T. H.,** A simple trimerization of 2,3-butanedione yielding a selective reagent for the modification of arginine in proteins, *J. Am. Chem. Soc.*, 90, 1664, 1968.
4. **Nakaya, K., Horinishi, H., and Shibata, K.,** States of amino acid residues in proteins. XIV. Glyoxal as a reagent for discrimination of arginine residues, *J. Biochem.*, 61, 345, 1967.
5. **Riordan, J. F., McElvany, K. D., and Borders, C. L., Jr.,** Arginyl residues: anion recognition sites in enzymes, *Science*, 195, 884, 1977.
6. **Patthy, L. and Thész, J.,** Origin of the selectivity of α-dicarbonyl reagents for arginyl residues of anion-binding sites, *Eur. J. Biochem.*, 105, 387, 1980.
7. **Takahashi, K.,** Specific modification of arginine residues in proteins with ninhydrin, *J. Biochem.*, 80, 1173, 1976.
8. **Patthy, L. and Smith, E. L.,** Reversible modification of arginine residues. Application to sequence studies by restriction of tryptic hydrolysis to lysine residues, *J. Biol. Chem.*, 250, 557, 1975.
9. **Chaplin, M. F.,** The use of ninhydrin as a reagent for the reversible modification of arginine residues in proteins, *Biochem. J.*, 155, 457, 1976.
10. **Hiraga, Y. and Kinoshita, T.,** Post-column derivatization of guanidino compounds in high-performance liquid chromatography using ninhydrin, *J. Chromatog.*, 226, 43, 1981.
11. **Jonas, A. and Weber, G.,** Presence of arginine residues at the strong, hydrophobic anion binding sites of bovine serum albumin, *Biochemistry*, 10, 1335, 1971.
12. **Glass, J. D. and Pelzig, M.,** Reversible modification of arginine residues with glyoxal, *Biochem. Biophys. Res. Commun.*, 81, 527, 1978.

13. **Pande, C. S., Pelzig, M., and Glass, J. D.,** Camphorquinone-10-sulfonic acid and derivatives: convenient reagents for reversible modification of arginine residues, *Proc. Natl. Acad. Sci. U.S.A.,* 77, 895, 1980.

14. **Rajagopalan, T. G., Stein, W. H., and Moore, S.,** The inactivation of pepsin by diazoacetylnorleucine methyl ester, *J. Biol. Chem.,* 241, 4295, 1966.

15. **Honegger, A., Hughes, G. J., and Wilson, K. J.,** Chemical modification of peptides by hydrazine, *Biochem. J.,* 199, 53, 1981.

16. **Sakaguchi, S.,** A new color reaction of protein and arginine, *J. Biochem.,* 5, 25, 1925.

17. **Izumi, Y.,** New Sakaguchi reaction, *Anal. Biochem.,* 10, 218, 1965.

18. **Izumi, Y.,** New Sakaguchi reaction. II, *Anal. Biochem.,* 12, 1, 1965.

19. **Enoch, H. G. and Strittmatter, P.,** Role of tyrosyl and arginyl residues in rat liver microsomal stearyl-coenzyme A desaturase, *Biochemistry,* 17, 4927, 1978.

20. **Smith, R. E. and MacQuarrie, R.,** A sensitive fluorometric method for the determination of arginine using 9,10-phenanthrenequinone, *Anal. Biochem.,* 90, 246, 1978.

21. **Fliss, H. and Viswanatha, T.,** 2,3-Butanedione as a photosensitizing agent: application to α-amino acids and α-chymotrypsin, *Can. J. Biochem.,* 57, 1267, 1979.

22. **Gripon, J.-C. and Hofmann, T.,** Inactivation of aspartyl proteinases by butane-2,3-dione. Modification of tryptophan and tyrosine residues and evidence against reaction of arginine residues, *Biochem. J.,* 193, 55, 1981.

23. **Mäkinen, K. K., Mäkinen, P.-L., Wilkes, S. H., Bayliss, M. E., and Prescott, J. M.,** Photochemical inactivation of *Aeromonas* aminopeptidase by 2,3-butanedione, *J. Biol. Chem.,* 257, 1765, 1982.

24. **Riley, H. A. and Gray, A. R.,** Phenylglyoxal, in *Organic Syntheses,* Collective Vol. 2, Blatt, A. H., Ed., John Wiley & Sons, New York, 1943, 509.

25. **Schloss, J. V., Norton, I. L., Stringer, C. D., and Hartman, F. C.,** Inactivation of ribulosebisphosphate carboxylase by modification of arginyl residues with phenylglyoxal, *Biochemistry,* 17, 5626, 1978.

26. **Augustus, B. W. and Hutchinson, D. W.,** The synthesis of phenyl[2-³H]glyoxal, *Biochem. J.,* 177, 377, 1979.

27. **Borders, C. L., Jr., Pearson, L. J., McLaughlin, A. E., Gustafson, M. E., Vasiloff, J., An, F. Y., and Morgan, D. J.,** 4-Hydroxy-3-nitrophenylglyoxal. A chromophoric reagent for arginyl residues in proteins, *Biochim. Biophys. Acta,* 568, 491, 1979.

28. **Yamasaki, R. B., Vega, A., and Feeney, R. E.,** Modification of available arginine residues in proteins by *p*-hydroxyphenylglyoxal, *Anal. Biochem.,* 109, 32, 1980.

29. **Cheung, S.-T. and Fonda, M. L.,** Reaction of phenylglyoxal with arginine. The effect of buffers and pH, *Biochem. Biophys. Res. Commun.,* 90, 940, 1979.

30. **Yamasaki, R. B., Shimer, D. A., and Feeney, R. E.,** Colorimetric determination of arginine residues in proteins by *p*-nitrophenylglyoxal, *Anal. Biochem.,* 111, 220, 1981.

31. **Steinbach, L. and Becker, E. I.,** A synthesis for β-aroylacrylic acids substituted with electron-withdrawing groups, *J. Am. Chem. Soc.,* 76, 5808, 1954.

32. **Branlant, G., Tritsch, D., and Biellmann, J.-F.,** Evidence for the presence of anion-recognition sites in pig-liver aldehyde reductase. Modification by phenylglyoxal and *p*-carboxyphenyl glyoxal of an arginyl residue located close to the substrate-binding site, *Eur. J. Biochem.,* 116, 505, 1981.

33. **Vanin, E. F., Burkhard, S. J., and Kaiser, I. I.,** *p*-Azidophenylglyoxal: a heterobifunctional photosensitive reagent, *FEBS Lett.,* 124, 89, 1981.

34. **Yankeelov, J. A., Jr.,** Modification of arginine in proteins by oligomers of 2,3-butanedione, *Biochemistry,* 9, 2433, 1970.

35. **Yankeelov, J. A., Jr. and Acree, D.,** Methylmaleic anhydride as a reversible blocking agent during specific arginine modification, *Biochem. Biophys. Res. Commun.,* 42, 886, 1971.

36. **Riordan, J. F.,** Functional arginyl residues in carboxypeptidase A. Modification with butanedione, *Biochemistry,* 12, 3915, 1973.

37. **Toi, K., Bynum, E., Norris, E., and Itano, H. A.,** Studies on the chemical modification of arginine. I. The reaction of 1,2-cyclohexanedione with arginine and arginyl residues of proteins, *J. Biol. Chem.,* 242, 1036, 1967.

38. **Patthy, L. and Smith, E. L.,** Identification of functional arginine residues in ribonuclease A and lysozyme, *J. Biol. Chem.,* 250, 565, 1975.

39. **Vallejos, R. H., Lescano, W. I. M., and Lucero, H. A.,** Involvement of an essential arginyl residue in the coupling activity of *Rhodospirillum rubrum* chromatophores, *Arch. Biochem. Biophys.,* 190, 578, 1978.

40. **Homyk, M. and Bragg, P. D.,** Steady-state kinetics and the inactivation by 2,3-butanedione of the energy-independent transhydrogenae of *Escherichia coli* cell membranes, *Biochim. Biophys. Acta,* 57, 201, 1979.

41. **Levy, H. M., Leber, P. D., and Ryan, E. M.,** Inactivation of myosin by 2,4-dinitrophenol and protection by adenosine triphosphate and other phosphate compounds, *J. Biol. Chem.,* 238, 3654, 1963.

42. **Bhagwat, A. S. and Ramakrishna, J.,** Essential histidine residues of ribulosebisphosphate carboxylase indicated by reaction with diethylpyrocarbonate and rose bengal, *Biochim. Biophys. Acta,* 662, 181, 1981.

43. **Philips, M., Pho, D. B., and Pradel, L.-A.,** An essential arginyl residue in yeast hexokinase, *Biochim. Biophys. Acta,* 566, 296, 1979.
44. **Mornet, D., Pantel, P., Audemard, E., and Kassab, R.,** Involvement of an arginyl residue in the catalytic activity of myosin heads, *Eur. J. Biochem.,* 100, 421, 1979.
45. **Cheung, S.-T. and Fonda, M. L.,** Kinetics of the inactivation of *Escherichia coli* glutamate apodecarboxylase by phenylglyoxal, *Arch. Biochem. Biophys.,* 198, 541, 1979.
46. **Davidson, W. S. and Flynn, T. G.,** A functional arginine residue in NADPH-dependent aldehyde reductase from pig kidney, *J. Biol. Chem.,* 254, 3724, 1979.
47. **Poulose, A. J. and Kolattukudy, P. E.,** Presence of one essential arginine that specifically binds the 2'-phosphate of NADPH on each of the ketoacyl reductase and enoyl reductase active sites of fatty acid synthetase, *Arch. Biochem. Biophys.,* 199, 457, 1980.
48. **Bond, M. W., Chiu, N. Y., and Cooperman, B. S.,** Identification of an arginine important for enzymatic activity within the covalent structure of yeast inorganic pyrophosphatase, *Biochemistry,* 19, 94, 1980.
49. **Mautner, H. G., Pakula, A. A., and Merrill, R. E.,** Evidence for presence of an arginine residue in the coenzyme A binding site of choline acetyltransferase, *Proc. Natl. Acad. Sci. U.S.A.,* 78, 7449, 1981.
50. **Hayman, S. and Colman, R. F.,** Effect of arginine modification on the catalytic activity and allosteric activation by adenosine diphosphate of the diphosphopyridine nucleotide specific isocitrate dehydrogenase of pig heart, *Biochemistry,* 17, 4161, 1978.
51. **Varimo, K. and Londesborough, S.,** Evidence for essential arginine in yeast adenylate cyclase, *FEBS Lett.,* 106, 153, 1979.
52. **Malinowski, D. P. and Fridovich, I.,** Chemical modification of arginine at the active site of the bovine erythrocyte superoxide dismutase, *Biochemistry,* 18, 5909, 1979.
53. **Schrijen, J. J., Luyben, W. A. H. M., DePont, J. J. H. M., and Bonting, S. L.,** Studies on (K$^+$ + H$^+$)-ATPase. I. Essential arginine residue in its substrate binding center, *Biochim. Biophys. Acta,* 597, 331, 1980.
54. **Fujioka, M. and Takata, Y.,** Role of arginine residue in saccharopine dehydrogenase (L-lysine forming) from baker's yeast, *Biochemistry,* 20, 468, 1981.
55. **Aurebekk, B. and Little, C.,** Functional arginine in phospholipase C of *Bacillus cereus, Int. J. Biochem.,* 8, 757, 1977.
56. **McTigue, J. J. and Van Etten, R. L.,** An essential arginine residue in human prostatic acid phosphatase, *Biochim. Biophys. Acta,* 523, 422, 1978.
57. **Ferti, C., Curti, B., Simonetta, M. P., Ronchi, S., Galliano, M., and Minchiotti, L.,** Reactivity of D-amino acid oxidase with 1,2-cyclohexanedione: evidence for one arginine in the substrate binding site, *Eur. J. Biochem.,* 119, 553, 1981.
58. **Belfort, M., Maley, G. F., and Maley, F.,** A single functional arginyl residue involved in the catalysis by *Lactobacillus casei* thymidylate synthetase, *Arch. Biochem. Biophys.,* 204, 340, 1980.

Chapter 2

CHEMICAL MODIFICATION OF TRYPTOPHAN

The specific chemical modification of tryptophan (Figure 1) in protein is one of the more challenging problems in protein chemistry. First, as will be apparent, the solvent conditions for providing specificity of modification are, in general, somewhat harsh. Secondly, there is the considerable possibility of either the concomitant or separate modification of a different amino acid residue. Thirdly, the analysis for the determination of the exact extent of modification requires a rigorous approach combining spectral analysis and amino acid analysis[1,2] after hydrolysis in a solvent which will not destroy tryptophan.

Treatment of tryptophan with hydrogen peroxide results in the oxidation of the indole ring.[3-6] Usually the reaction is performed at alkaline pH (1.0 M sodium bicarbonate, pH 8.4) with the H_2O_2 dioxane mixture prepared as described by Hachimori and co-workers.[3] The loss of tryptophan is monitored by the change in absorbance at 280 nm.[3,5,6] The difference in the molar extinction coefficient between tryptophan and the fully oxidized derivative is 3490 M^{-1} cm^{-1}.

The reaction of N-bromosuccinimide (NBS) with protein has been studied in some detail (Figure 2). This reagent was introduced for use in protein chemistry in 1958.[7] The early work with this reagent was summarized in 1967.[8-10]

The use of oxidation with N-bromosuccinimide to determine the tryptophan content of proteins can be of some value. One adds small increments of a freshly prepared solution of N-bromosuccinimide until there is no further decrease in absorbance at 280 nm. The change in the molar extinction coefficient of tryptophan on conversion to the oxindole derivative is taken to be 4×10^3 M^{-1} cm^{-1}.[11] It has been our experience that one must either perform the reaction in 8.0 M urea (pH adjusted to 4.0) or with the reduced, carboxymethylated derivative.[12] The spectra must be obtained as soon as possible after the addition of the N-bromosuccinimide since, unless the excess reagent and low molecular weight products of the reaction are rapidly removed, there is a reversal of the decrease in absorbance.[13] This is not a trivial consideration since there is at least one study[14] where there is a real difference in the extent of modification as determined by spectroscopy or amino acid analysis. The rigorous evaluation[13] of the reaction of N-bromosuccinimide with model tryptophanyl and tyrosyl compounds reported from Keitaro Hiromi's laboratory provides considerable insight into the problems to be encountered with the study of intact proteins. Figure 3 shows the changes in the UV spectrum of N-acetyl-tryptophan ethyl ester (ATEE) upon reaction with N-bromosuccinimide. These spectra were obtained within 5 min after the initiation of the reaction. At ratios of N-bromosuccinimide to N-acetyl-tryptophan ethyl ester of greater than 2 there is an apparent reversal of the decrease in absorbance at 280 nm as shown in Figure 4. Figure 5 shows the spectral changes occurring upon the reaction of N-bromosuccinimide with N-acetyl-tryptophan ethyl ester as a function of time and molar excess of N-bromosuccinimide. The maximal decrease in absorbance occurs at a ratio of N-bromosuccinimide to tryptophan of 2. If the data are obtained by stopped-flow spectroscopy, the molar excess of N-bromosuccinimide does not have an effect on the maximum decrease observed, but when the spectrum is obtained 5 min after the initiation of the reaction, there is a decrease in the observed magnitude of change in absorbance at 280 nm. The evaluation of spectral changes in a protein is further complicated by the reaction of N-bromosuccinimide with tyrosine. This is demonstrated in Figure 6 which shows the spectral changes occurring as a result of reaction of N-bromosuccinimide with N-acetyl-tyrosine ethyl ester. Here an increase in absorbance at 280 nm can be observed. Figure 7 shows that the increase in absorbance of N-acetyl-tyrosine ethyl ester on reaction with N-bromosuccinimide is dependent on the molar excess of N-bromosuccinimide. The use of this procedure for the analysis of tryptophan

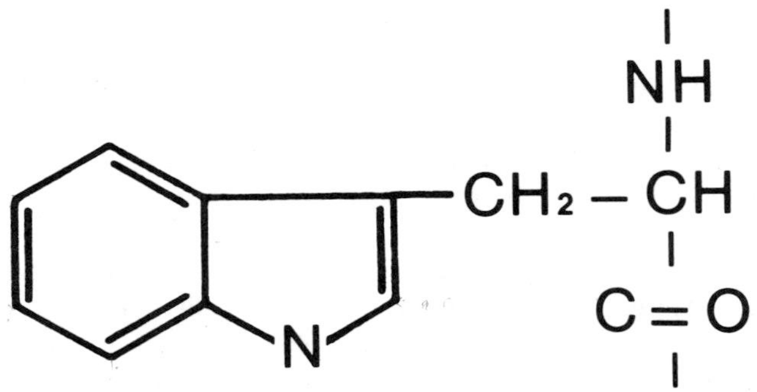

FIGURE 1.   The structure of tryptophan.

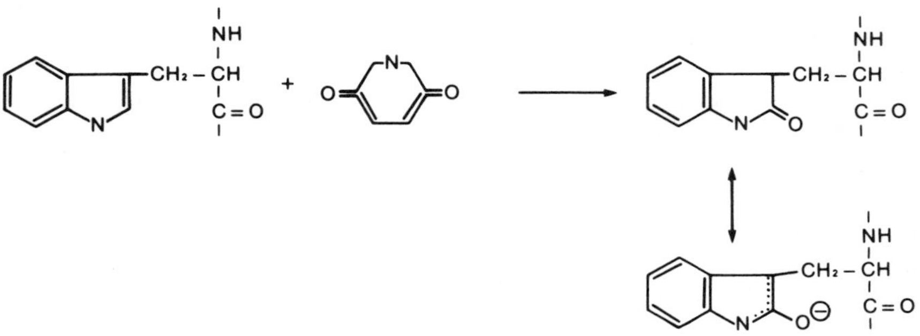

FIGURE 2.   The reaction of tryptophan with *N*-bromosuccinimide.

content in proteins has been largely supplanted by ion-exchange analysis following modified hydrolytic procedures.[1,2,14]

The primary use of the *N*-bromosuccinimide modification of proteins during the past decade has been in studies on the effect of such modification on biological (catalytic) activity. In general, the modification reaction is performed in 0.1 *M* sodium acetate, pH 4 to 5. The *N*-bromosuccinimide should be recrystallized from water before use. The presence of halides such as chloride or bromide in the solvent must be avoided since the addition of *N*-bromosuccinimide will oxidize these ions to the elemental form with disastrous and irreproducible effects on the proteins under study. In general, a twofold molar excess of *N*-bromosuccinimide per mole of tryptophan is necessary to achieve modification. Daniel and Trowbridge[15] found that (at pH 4.0) the reaction of *N*-bromosuccinimide with acetyl-L-tryptophan ethyl ester required 1.5 mol of *N*-bromosuccinimide per mole of the acetyl-L-tryptophan ethyl ester, while trypsinogen required 2.0 to 2.3 mol *N*-bromosuccinimide per mole of tryptophan oxidized, and trypsin required 1.5 to 2.0 mol *N*-bromosuccinimide per mole tryptophan oxidized. An interesting phenomenon was reported by Freisheim and Huennekens.[16] At pH 4.0, only tryptophan in dihydrofolate reductase reacts with NBS while at pH 6.0, a sulfhydryl group apparently is preferentially oxidized by the reagent prior to the reaction of tryptophan. This observation was pursued in greater detail by Freisheim et al.[17] Figure 8 shows the *observed* relationship between the decrease in absorbance at 280 nm and the catalytic activity of dihydrofolate reductase as a function of the molar excess of *N*-bromosuccinimide at pH 6.5 (0.1 *M* phosphate). The initial increase in activity reflects oxidation of a cysteinyl residue while the decrease in activity seen at higher molar excess of reagent appears to be related to the oxidation of tryptophanyl residues in the protein

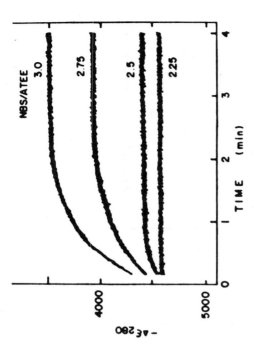

FIGURE 4. Time course of the difference absorbance change of *N*-acetyltryptophan ethyl ester at 280 nm caused by *N*-bromosuccinimide. *N*-acetyltryptophan ethyl ester, 49 μ*M*; 0.1 *M* acetate buffer, pH 4.5; 25°C. The numbers in the figure indicate the molar ratios of *N*-bromosuccinimide to *N*-acetyltryptophan ethyl ester. (From Ohnishi, M., Kawagishi, T., Abe, T., and Hiromi, K., *J. Biochem.*, 87, 273, 1980. With permission.)

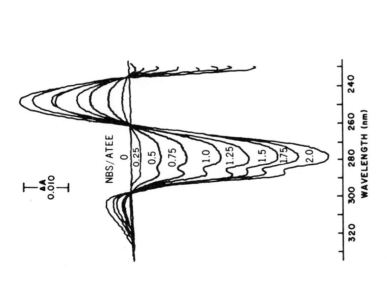

· FIGURE 3. Difference UV absorption spectra of *N*-acetyltryptophan ethyl ester caused by reaction with *N*-bromosuccinimide. *N*-acetyltryptophan ethyl ester 19 μ*M*, numbers indicate the molar ratios of *N*-bromosuccinimide to *N*-acetyltryptophan ethyl ester. The reactions were performed in 0.1 *M* acetate buffer, pH 4.5 at 25°C. The difference spectra were obtained within 5 min of the start of the reaction. (From Ohnishi, M., Kawagishi, T., Abe, T., and Hiromi, K., *J. Biochem.*, 87, 273, 1980. With permission.)

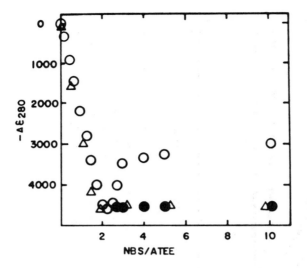

FIGURE 5. Dependence of the difference UV absorption change (decrease) at 280 nm of *N*-acetyltryptophan ethyl ester (ATEE) on time and molar excess of *N*-bromosuccinimide (NBS). The open circles show the difference at 5 min after the addition of *N*-bromosuccinimide and the closed circles the change at 0 min (obtained by extrapolation of the time curves). The triangles represent data obtained by the stopped-flow method. The concentration of *N*-acetyltryptophan ethyl ester was 49 μM in 0.1 M acetate buffer, pH 4.5, at 25°C. (From Ohnishi, M., Kawagishi, T., Abe, T., and Hiromi, K., *J. Biochem.*, 87, 273, 1980. With permission.)

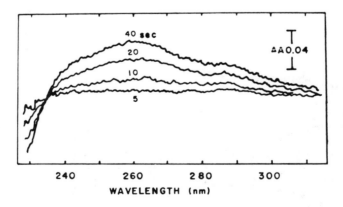

FIGURE 6. The difference UV absorption spectra of *N*-acetyltyrosine ethyl ester on reaction with *N*-bromosuccinimide. *N*-acetyltyrosine ethyl ester, 25 μM; *N*-bromosuccinimide, 100 μM; 0.1 M acetate buffer, pH 4.5; 25°C. The data were obtained with a stopped-flow spectrophotomer in a rapid scanning mode. The spectra were recorded at 5, 10, 20, and 40 sec after the start of the reaction. A spectrum was obtained within 10 msec (scan speed). (From Ohnishi, M., Kawagishi, T., Abe, T., and Hiromi, K., *J. Biochem.*, 87, 273, 1980. With permission.)

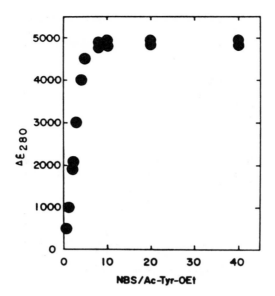

FIGURE 7. Dependence of the difference UV absorption change (increase) of *N*-acetyl-tyrosine ethyl ester (Ac-Tyr-OET) caused by reaction with *N*-bromosuccinimide on the *N*-bromosuccinimide (NBS)/*N*-acetyltyrosine ethyl ester ratio. *N*-acetyltyrosine ethyl ester, 25 μ*M*; 0.1 *M* acetate, pH 4.5; 25°C. The spectra were obtained 5 min after the start of the reaction. (From Ohnishi, M., Kawagishi, T., Abe, T., and Hiromi, K., *J. Biochem.*, 87, 273, 1980. With permission.)

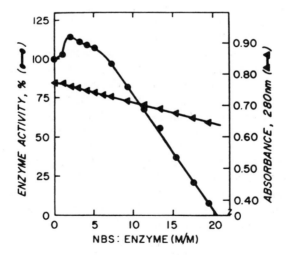

FIGURE 8. Activity and UV absorbance changes of dihydrofolate reductase as a function of the molar excess of *N*-bromosuccinimide. The enzyme concentration was 9.8 μ*M* in 0.05 *M* potassium phosphate, pH 6.5, at 25°C. The maximum changes in enzyme activity or absorbance at 280 nm occurred in the first 2 to 3 min following the addition of *N*-bromo-succinimide. Enzyme activity is expressed as percent of an untreated control. (From Warwick, P. E., D'Souza, L., and Freisheim, J. H., *Biochemistry*, 11, 3775, 1972. With permission.)

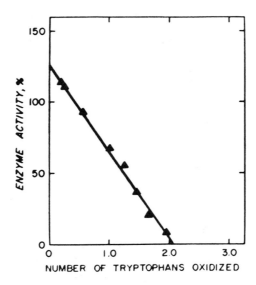

FIGURE 9.   Stoichiometry for the inactivation of dihydrofolate reductase by *N*-bromosuccinimide. Shown is the activity of dihydrofolate reductase as a function of the number of tryptophanyl residues oxidized. The experimental conditions were as described in Figure 8. (From Warwick, P. E., D'Souza, L., and Freisheim, J. H., *Biochemistry*, 11, 3775, 1972. With permission.)

(Figure 9). Poulos and Price have reported on the reaction of a tryptophanyl residue in bovine pancreatic DNAse with *N*-bromosuccinimide.[18] This study was of some interest in that prior reaction of the DNAse with another "tryptophan" reagent, 2-hydroxy-5-nitrobenzyl bromide, modified a different residue from the one modified by *N*-bromosuccinimide. These investigators used spectral analysis to determine the extent of tryptophan modification. Subsequent studies from another laboratory[14] on the modification of DNAse with *N*-bromosuccinimide suggested that apparently 2 mol of tryptophan are modified per mole of enzyme at 100% inactivation with a sixfold molar excess of *N*-bromosuccinimide in 0.01 M CaCl$_2$ at pH 4.0. Using amino acid analysis (after hydrolysis in 6 N HCl containing mercaptoacetic acid, phenol and 3-(2-aminoethyl) indole for 24 hr at 110°C), these investigators showed that all three tryptophanyl residues are modified under the above experimental conditions. A study on the reaction of *N*-bromosuccinimide with relaxin is interesting in that the modification of the first mole of tryptophan does not result in loss of biological activity while reaction of the second residue is associated with the loss of biological activity.[19] Figure 10 shows a typical absorbance spectrum for the reaction of *N*-bromosuccinimide with the protein as reported by these investigators. The study on the modification of tryptophan in galactose oxidase[20] is worth comment in that these investigators report the amino acid composition of the modified protein after hydrolysis in 3 N *p*-toluenesulfonic acid. There was excellent agreement between the extent of tryptophan modification as judged by direct amino acid analysis and the value observed by spectral analyses. This study shows one of the consequences of the conversion of tryptophan from the indole to the oxindole. Tryptophan is responsible for the majority of the innate fluorescence of proteins and oxidation by *N*-bromosuccinimide obviates this property as shown in Figure 11. Although the reaction between *N*-bromosuccinimide and tryptophan residues in protein is quite rapid, Fujimori and co-workers,[21] using stopped-flow kinetics, were able to determine kinetically different tryptophanyl residues in *Bacillus subtilis* α-amylase. Figure 12 shows the change in the absorbance spectrum of the protein on reaction with *N*-bromosuccinimide, obtained 5 min after

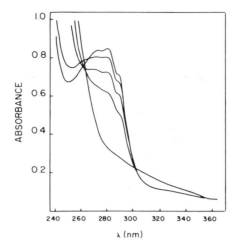

FIGURE 10. The stepwise oxidation of tryptophan in purified relaxin. A salt-free sample of relaxin (1.5 mg) was dissolved in exactly 1 mℓ 0.2 *M* sodium acetate, pH 4.7. The oxidation of tryptophan was accomplished by adding 2 μℓ portions of 10 m*M* *N*-bromosuccinimide. The decrease in absorbance at 280 nm as a function of *N*-bromosuccinimide concentration can be evaluated without complication because of the lack of tyrosine in relaxin. The upper line represents the spectrum of unmodified relaxin; the lowest line (at 280 nm) represents the spectrum of the completely oxidized relaxin. (From Schwabe, C. and Braddon, S. A., *Biochem. Biophys. Res. Commun.*, 68, 1126, 1976. With permission.)

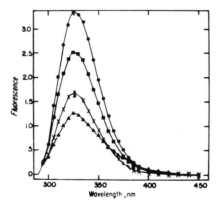

FIGURE 11. Corrected fluorescence emission spectra of deoxygenated solutions of galactose oxidase: unmodified (●); with 0.85 oxidized tryptophans (■); with 2.0 oxidized tryptophans (×); and with 3.0 oxidized tryptophans (▲). The spectra were recorded in 100 m*M* sodium acetate, pH 4.15, after the modification with *N*-bromosuccinimide was performed in 5 m*M* sodium acetate, pH 4.15. The protein concentration was 0.14 mg/mℓ. The error bars represent the standard deviation of time-averaged recordings. (From Kosman, D. J., Ettinger, M. J., Bereman, R. D., and Giordano, R. S., *Biochemistry*, 16, 1597, 1977. With permission.)

the addition of reagent. Figure 13 compares the extent of reaction after 5 min with that obtained using stopped-flow techniques (Figure 14). Four of the eleven tryptophan residues were modified at the maximum extent of reaction but one of these clearly reacted more rapidly than the other residues. Similar results were obtained when changes in the intrinsic fluorescence of the protein were used to monitor the reaction as shown in Figure 15. These

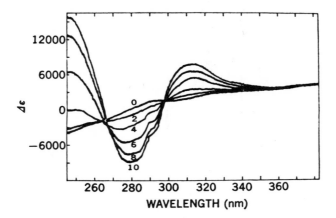

FIGURE 12.   Changes in the UV absorption spectrum of *Bacillus subtilis* α-amylase on reaction with *N*-bromosuccinimide. Shown are the difference spectra of the enzyme caused by modification with *N*-bromosuccinimide at pH 7.0 (0.01 *M* phosphate buffer) at 25°C. The numbers in the figure indicate the molar ratio of *N*-bromosuccinimide to the enzyme (0, base line, no reagent added). The spectra were taken at 5 min after mixing. The enzyme concentration was 2.8 μ*M*. Δ E, Difference absorbance per mole of the enzyme. (From Fujimori, H., Ohnishi, M., and Hiromi, K., *J. Biochem.*, 83, 1503, 1978. With permission.)

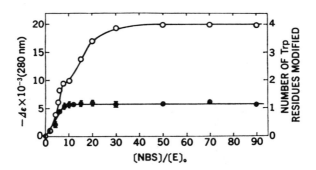

FIGURE 13.   Spectrophotometric titration of *Bacillus subtilis* α-amylase with *N*-bromosuccinimide (NBS) at pH 7.0. The enzyme concentration was fixed at 2.8 μ*M*. ○, The value of $-\Delta E_{280}$ was measured with a spectrophotometer at 5 min after mixing the enzyme and *N*-bromosuccinimide solutions in a quartz cuvette. ●, The most rapid decrease in absorbance at 280 nm observed by the stopped-flow method within 0.3 sec. The number of modified tryptophan residues was calculated from ΔE_{280} using a molar difference absorption per mol of tryptophan residue at 280 nm(ΔE_{280}) of 5,000 cm^{-1}. (From Fujimori, H., Ohnishi, M., and Hiromi, K., *J. Biochem*, 83, 1503, 1978. With permission.)

investigators were able to determine a second-order rate constant of $3.5 \times 10^5 \, M^{-1} \, \text{sec}^{-1}$ for the tryptophanyl residue reacting most rapidly.

There are several other facets of the use of *N*-bromosuccinimide for the modification of tryptophanyl residue in proteins which should be considered. The use of the reagent at mildly

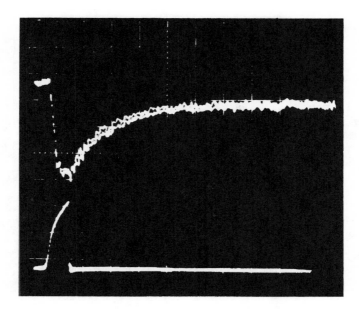

FIGURE 14. Oscilloscope traces obtained by the stopped-flow method for the reaction of *Bacillus subtilis* α-amylase (2.8 μ*M*) with *N*-bromosuccinimide (22.4 μ*M*) in 0.01 *M* sodium phosphate, pH 7.0, at 25°C. The vertical axis represents the transmittance (0.005 O.D. per major division). The horizontal axis is a time scale corresponding to 50 msec per major division. The lower trace is the flow velocity curve. Optical path was 2 mm. (From Fujimori, H., Ohnishi, M., and Hiromi, K., *J. Biochem.*, 83, 1503, 1978. With permission.)

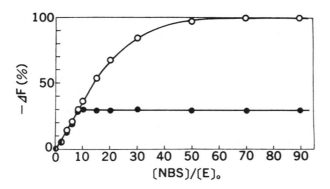

FIGURE 15. The fluorescence titration curve of *Bacillus subtilis* α-amylase with *N*-bromosuccinimide (NBS) at pH 7.0. The enzyme concentration was fixed at 2.8 μ*M* in 0.01 *M* sodium phosphate, pH 7.0. The decrease in fluorescence intensity at 340 nm($-\Delta F$) of the enzyme excited at 280 nm caused by the addition of *N*-bromosuccinimide is expressed in terms of the percentage fluorescence intensity change with respect to the fluorescence intensity of the native enzyme and plotted against the molar ratio of *N*-bromosuccinimide to the enzyme. ●, The most rapid decrease in fluorescence intensity observed by the stopped-flow method within 0.3 sec. ○, The value obtained with a spectrofluorometer at 5 min after mixing. (From Fujimori, H., Ohnishi, M., and Hiromi, K., *J. Biochem.*, 83, 1503, 1978. With permission.)

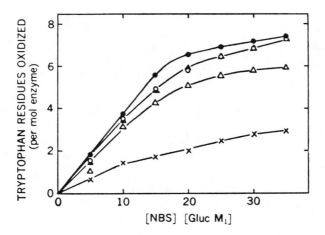

FIGURE 16.   *N*-bromosuccinimide oxidation of a glucoamylase (Gluc M$_1$) as a function of pH. *N*-Bromosuccinimide (6.6 m*M*) was added in 1 to 10 μ*ℓ* portions at 5 min intervals to 1 m*ℓ* of 6.67 μ*M* Gluc M$_1$ in 0.1 *M* acetate buffer at pH 4.0(●), pH 4.5(○), pH 5.0(▲), pH 6.0(△), and pH 7.0(×). The decrease in absorbance at 280 nm was measured after each addition of *N*-bromosuccinimide at 25°C. The amount of tryptophan residues oxidized was calculated according to the method of Spande and Witkop. (From Inokuchi, N., Takahashi, T., Yoshimoto, A., and Irie, M., *J. Biochem.*, 91, 1661, 1982. With permission.)

acidic pH has been mentioned above. Not only does increasing pH decrease specificity in terms of reaction with amino acid residues other than tryptophan but there is a decrease in the modification of tryptophan. This is shown by the studies[22] on the modification of a glucoamylase from *Aspergillus saitoi*. As shown in Figure 16 there is a modest decrease in modification as the pH is increased from 4.0 to 6.0 with a dramatic decrease at pH 7.0. At pH values close to neutrality there is the increased possibility of modification of amino residues other than tryptophan. In studies[23] on the reaction on *Escherichia coli* lac repressor protein with *N*-bromosuccinimide at pH 7.8 (1.0 *M* Tris), cysteine was modified as readily as tryptophan with lesser modification of methionine and tyrosine (Figure 17). The use of excessive *N*-bromosuccinimide should also be avoided as shown by studies[24] on amino acylase[24] (Figure 18). In general, modification should occur at a 4 to 6 molar excess (with respect to total tryptophan) of *N*-bromosuccinimide. The use of *N*-bromosuccinimide in the study of proteins is summarized in Table 1.

The reaction of *N*-bromosuccinimide with proteins can also result in the cleavage of peptide bonds at tryptophan, tyrosine, and histidine.[25] Thus, the careful investigator will also evaluate the integrity of the polypeptide chain(s) of the protein of interest. Whereas peptide bond cleavage is usually an unwanted side reaction, Feldhoff and Peters[26] have devised a procedure which has enhanced specificity for tryptophan. Their procedure uses 8.0 *M* urea, 2.0 *M* acetic acid as the solvent with a 20-fold molar excess of *N*-bromosuc-cinimide. Their approach offers at least two advantages; first, the protein is denatured so that all residues should be equally available and, second, the *N*-bromosuccinimide reacts with urea to yield *N*-bromourea, a less severe oxidizing agent which should have increased specificity for tryptophanyl residues. The use of *N*-chlorosuccinimide for peptide bond cleavage of tryptophanyl residues has also been discussed.[27]

The conversion of tryptophanyl residues to 1-formyltryptophanyl residues has been re-ported. The reaction conditions are somewhat harsh, but the procedure is reversible (Figure

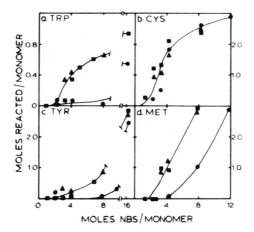

FIGURE 17.   The modification of *Escherichia coli* lac repressor protein with *N*-bromosuc-
cinimide. Shown are the moles of amino acid reacted with *N*-bromosuccinimide per monomer
of repressor protein. The modification was performed in 1.0 *M* Tris/Cl, pH 7.8, for 15 min
at ambient temperature in the dark. The reactions were terminated by the addition of dithiothreitol and the modified protein preparations dialyzed vs. water for subsequent analysis.
The amount of tryptophan, tyrosine, and methionine were determined by amino acid analysis
after hydrolysis in methanesulfonic acid. Cysteine was determined by titration with 2-chlo-
romercuri-4-nitrophenol in 8 *M* urea. ■——■, Repressor reacted with *N*-bromosuccinimide
alone; ●——●, repressor reacted with *N*-bromosuccinimide in the presence of isopropyl-l-
thio-β-D-galactoside; ▲——▲, repressor reacted with *N*-bromosuccinimide in the presence
of *o*-nitrophenyl-β-D-fucoside. (From O'Gorman, R. B. and Matthews, K. S., *J. Biol. Chem.*,
252, 3565, 1977. With permission.)

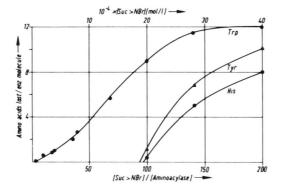

FIGURE 18.   The modification of aminoacylase with *N*-
bromosuccinimide. Shown is the modification of tryptophan (●——●), tyrosine (▲——▲), and histidine (○—
——○) with increasing amounts of *N*-bromosuccinimide
(Suc>NBr) in 0.1 *M* acetate buffer, pH 5.0, containing
0.1 *M* urea at 25°C. (From Kördel, W. and Schneider,
F., *Hoppe-Seyler's Z. Physiol. Chem.*, 357, 1109, 1976.
With permission.)

19) and should prove quite useful for small peptides and has been applied to at least two
proteins. Coletti-Previero and co-workers[28] have successfully applied this procedure to bovine
pancreatic trypsin. Trypsin was dissolved in formic acid saturated with HCl at a concentration

Table 1
EXAMPLES OF THE MODIFICATION OF PROTEINS WITH
N-BROMOSUCCINIMIDE

Protein	Solvent	Molar excess[a]	Extent of modification	Ref.
Trypsinogen	pH 7.0[b]	1—4	1—2	1
Trypsin	pH 4.0[b]	1—4	1—2	1
Dihydrofolate reductase	0.1 M sodium phosphate, pH 6.0	15	2.0	2
		15	2.7	2
	0.1 M sodium acetate, pH 4.0 0.13 M sodium acetate formate, 5.3 M urea, pH 4.0	12	3.8	2
Bovine pancreatic DNAse	0.1 M sodium acetate, pH 4.0 containing 0.033 M CaCl₂	6	1.0	3
Bovine pancreatic DNAse	pH 4.0, 0.010 M CaCl₂	1—6	3[c]	4[c]
Dihydrofolate reductase	0.05 M potassium phosphate, pH 6.5	20	2.0	5
Pyrocatechase (*B. fuscum*)[d]	0.1 M phosphate, pH 7.0	—	2	6
Relaxin	0.2 M sodium acetate, pH 4.7	—		7
Rhodopsin	0.1 M Tris acetate, pH 7.4 containing 1% emulphogene	50	6	8
		100	9	8
Pig kidney amino acylase	0.1 M sodium acetate, pH 5.0—1.0 M urea	50	6	9
Galactose oxidase	0.005 M sodium acetate, pH 4.15	7	2	10
Bovine thrombin	0.1 Sodium acetate, pH 4.0	1	0.5	11
		2	1.1	11
Papain[e]	0.05 M sodium acetate, pH 4.75	6[f]	1.4	12
Lac repressor protein	1.0 M Tris HCl, pH 7.8	8	0.7[g]	13
α-Mannosidase (*P. vulgaris*)	1.0 M sodium acetate, pH 4.0	35	10	14
α-Amylase (*B. subtilis*)	0.01 M sodium phosphate, pH 7.0	8	2	15
		50	4	15
Dihydrofolate reductase	0.015 M Bis Tris HCl, pH 6.5			16
	0.5 M KCl	4	1.2	

[a]   Reagent to protein.
[b]   pH maintained at 4.0 by addition of KOH.
[c]   Spectral analysis suggested 2 mol tryptophan oxidized while amino acid analysis demonstrates that all three tryptophan residues modified.
[d]   Thiophenylated apoenzyme [apoenzyme modified with 5,5'-dithiobis (2-nitrobenzoic acid)].
[e]   Not activated.
[f]   Also modified tyrosine at this concentration.
[g]   Also had substantial modification of tyrosine, cysteine, and methionine.

References for Table 1

1. **Daniel, V. W., III and Trowbridge, C. G.,** The effect of *N*-bromosuccinimide upon trypsinogen activation and trypsin catalysis, *Arch. Biochem. Biophys.,* 134, 506, 1969.
2. **Freisheim, J. H. and Huennekens, F. M.,** Effect of *N*-bromosuccinimide on dihydrofolate reductase, *Biochemistry,* 8, 2271, 1969.
3. **Poulos, T. L. and Price, P. A.,** The identification of a tryptophan residue essential to the catalytic activity of bovine pancreatic deoxyribonuclease, *J. Biol. Chem.,* 246, 4041, 1971.
4. **Sartin, J. L., Hugli, T. E., and Liao, T.-H.,** Reactivity of the tryptophan residues in bovine pancreatic deoxyribonuclease with *N*-bromosuccinimide, *J. Biol. Chem.,* 255, 8633, 1980.

Table 1 (continued)

5. **Warwick, P. E., D'Souza, L., and Freisheim, J. H.,** Role of tryptophan in dihydrofolate reductase, *Biochemistry,* 11, 3775, 1972.
6. **Nagami, K.,** The participation of a tryptophan residue in the binding of ferric iron to pyrocatechase, *Biochem. Biophys. Res. Commun.,* 51, 364, 1973.
7. **Schwabe, C. and Braddon, S. A.,** Evidence for the essential tryptophan residue at the active site of relaxin, *Biochem. Biophys. Res. Commun.,* 68, 1126, 1976.
8. **Cooper, A. and Hogan, M. E.,** Reactivity of tryptophans in rhodopsin, *Biochem. Biophys. Res. Commun.,* 68, 178, 1976.
9. **Kördel, W. and Schneider, F.,** Chemical modification of two tryptophan residues abolishes the catalytic activity of aminoacylase, *Hoppe Seyler's Z. Physiol. Chem.,* 357, 1109, 1976.
10. **Kosman, D. J., Ettinger, M. J., Bereman, R. D., and Giordano, R. S.,** Role of tryptophan in the spectral and catalytic properties of the copper enzyme, galactose oxidase, *Biochemistry,* 16, 1597, 1977.
11. **Uhteg, L. C. and Lundblad, R. L.,** The modification of tryptophan in bovine thrombin, *Biochim. Biophys. Acta,* 491, 551, 1977.
12. **Glick, B. R. and Brubacher, L. S.,** The chemical and kinetic consequences of the modification of papain by *N*-bromosuccinimide, *Can. J. Biochem.,* 55, 424, 1977.
13. **O'Gorman, R. B. and Matthews, K. S.,** *N*-bromosuccinimide modification of *lac* repressor protein, *J. Biol. Chem.,* 252, 3565, 1977.
14. **Paus, E.,** The chemical modification of tryptophan residues of α-mannosidase from *Phaseolus vulgaris,* *Biochim. Biophys. Acta,* 533, 446, 1978.
15. **Fujimori, H., Ohnishi, M., and Hiromi, K.,** Tryptophan residues of saccharifying α-amylase from *Bacillus subtilis.* A kinetic discrimination of states of tryptophan residues using *N*-bromosuccinimide, *J. Biochem.,* 83, 1503, 1978.
16. **Thomson, J. W., Roberts, G. C. K., and Burgen, A. S. V.,** The effects of modification with *N*-bromosuccinimide on the binding of ligands to dihydrofolate reductase, *Biochem. J.,* 187, 501, 1980.

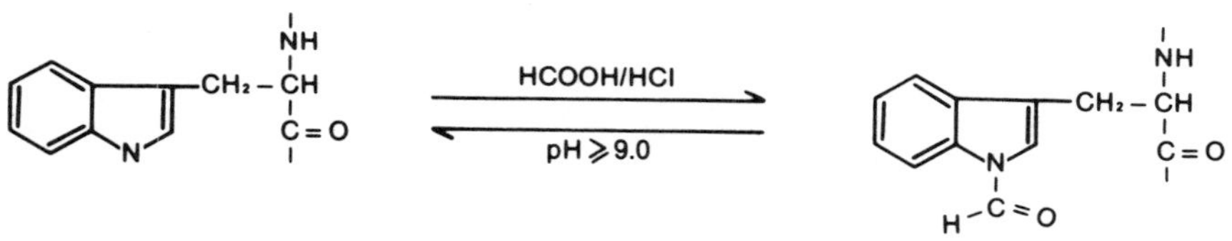

FIGURE 19.   A scheme for the reversible formylation of tryptophan residues.

of 2.5 mg/mℓ at 20°C. The formylation reaction is associated with an increase in absorbance at 298 nm[29] (Figure 20). It therefore is possible to follow the reaction spectrophotometrically. The reaction is judged complete when there is no further increase in absorbance at 298 nm. The above reaction with trypsin was complete after an incubation period of 1 hr. The solvent was partially removed *in vacuo* over KOH pellets followed by lyophilization. The formyl-tryptophan derivative is stable at pH 8.0. This is shown in Figure 21 where there is an increase in esterase activity which is lost as a result of denaturation during the modification reaction. Note that there is not recovery of proteinase activity under these conditions. At pH 9.5 (pH-Stat), conversion back to tryptophan is complete after 200 min incubation at 20°C as shown in Figure 22. The relationship between the reversal of the formylation reaction and the recovery of proteinase activity is shown in Figure 23. From these data and those shown in Figure 22 these investigators concluded that there is not preferential deformylation of any of the four tryptophanyl residues in trypsin and that the recovery of proteolytic activity is associated with the deformylation of three of the formyl-tryptophanyl residues modified in trypsin. Holmgren has successfully applied this procedure to thioredoxin.[30] A procedure for the acylation of the carbon at position 2 on the indole ring has also been reported.[31]

One of the most useful modification procedures for tryptophanyl residues in proteins involves the use of 2-hydroxy-5-nitrobenzyl bromide and its various derivatives. 2-Hydroxy-5-nitrobenzyl bromide, frequently referred to as Koshland's reagent, was introduced by Koshland and co-workers.[32,33] Barman and Koshland[34] have reported the use of 2-hydroxy-

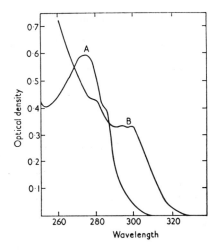

FIGURE 20.   Changes in the UV absorption spectrum of trypsin occurring as a result of formylation of tryptophan residues. Shown is the spectrum of trypsin (18 μM in 8.0 M urea, pH 4.0) before (curve A) and after (curve B) 1-formylation of tryptophyl residues. Formylation of trypsin was accomplished by dissolving trypsin in formic acid saturated with gaseous HCl at 20°C (2.5 mg/mℓ). At suitable time intervals, 0.4 mℓ samples of the solution were diluted with 2 mℓ of 8 M urea, pH 4.0 for recording the UV spectra. When the maximum increase in absorbance at 298 nm was reached (about 60 min), the solvent was partially removed under vacuum over KOH pellets for 15 min in order to eliminate most of the HCl and the sample was subsequently lyophilized. (From Coletti-Previero, M.-A., Previero, A., and Zuckerkandl, E., *J. Mol. Biol.,* 39, 493, 1969. With permission.)

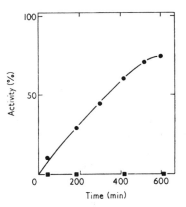

FIGURE 21.   Recovery of enzymic activity during renaturation by incubation of formylated trypsin at pH 8.0 at 20°C.●——●, Esterase activity (BzArgOEt; ■——■, protease activity (BzArgpNA, casein). (From Coletti-Previero, M. A., Previero, A., and Zuckerkandl, E., *J. Mol. Biol.,* 39, 493, 1969. With permission.)

5-nitrobenzyl bromide for the quantitative determination of tryptophanyl residues in proteins. Although this approach to the quantitative determination of tryptophanyl residues in proteins has been largely replaced by the development of new methods for the hydrolysis of proteins, it can still be useful in certain instances. For this procedure the sample is incubated for 16 to 20 hr at 37°C in 1.0 mℓ 10 M urea (the urea should be recrystallized (EtOH/H$_2$O) prior to use), pH 2.7 (pH adjusted with concentrated HCl). This solution is cooled to ambient

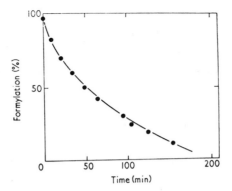

FIGURE 22.   Deformylation of tryptophanyl residues in trypsin by incubation at alkaline pH. Shown is the hydrolysis of 1-formyltryptophan in trypsin during incubation at pH 9.5, as measured spectrophotometrically at 298 nm(ϵ = 4880) at 20°C. The solution of formylated trypsin after incubation at pH 8.0 (see Figure 21) was allowed to stand in the reaction chamber of a pH Stat at pH 9.5, 25°C and maintained at this value by the addition of 0.1 M NaOH. (From Coletti-Previero, M.-A., Previero, A., and Zuckerkandl, E., *J. Mol. Biol.*, 39, 493, 1969. With permission.)

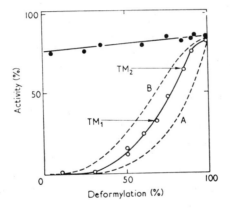

FIGURE 23.   Recovery of protease activity during deformylation of 1-formyl trypsin at pH 9.5 at 20°C. ●———●, Esterase activity (BzArgOEt); ○———○, protease activity (BzArgpNA and casein). Theoretical curve A is based on the assumption that the deformylation of all four tryptophan residues per molecule is required for the restoration of protease activity. Theoretical curve B is based on the assumption that the deformylation of any three tryptophanyl residues per molecule will lead to the restoration of protease activity. TM₁ and TM₂ indicate samples removed during the process of deformylation for subsequent analysis. (From Coletti-Previero, M.-A., Previero, A., and Zuckerkandl, E., *J. Mol. Biol.*, 39, 493, 1969. With permission.)

temperature and approximately 5.0 mg of 2-hydroxy-5-nitrobenzyl bromide (in 0.1 mℓ acetone) is added followed by vigorous stirring (we have found the Pierce Reacti-Vials® very useful for this purpose). Occasionally a precipitate of 2-hydroxy-5-nitrobenzyl alcohol (the hydrolytic product of 2-hydroxy-5-nitrobenzyl bromide) forms which can be removed by centrifugation. The labeled protein is obtained free of reagent by gel filtration. This step is generally performed under acidic condition (e.g., 0.18 M acetic acid, 10% acetic acid or

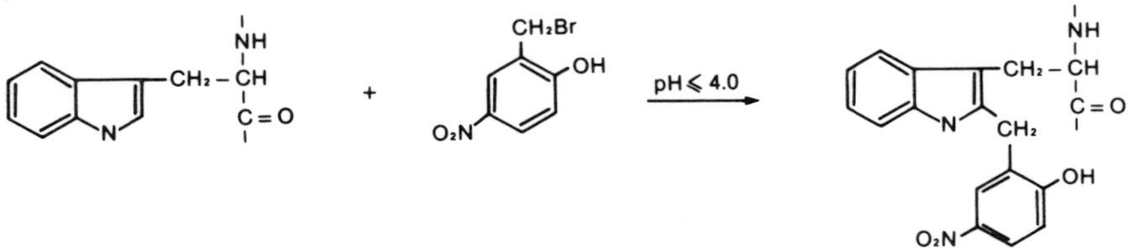

FIGURE 24.   A scheme for the reaction of 2-hydroxy-5-nitrobenzyl bromide with tryptophan.

10% formic acid).* Depending upon the protein under study, it might be necessary to perform this step in 10 M urea (pH 2.7) to maintain the solubility of the modified protein. A portion of the modified protein is taken to a pH greater than 12 with NaOH. The extent of incorporation is determined at 410 nm using an extinction coefficient of 18,000 M^{-1} cm^{-1}. It is necessary to determine the concentration of protein by a technique other than absorbance at 280 nm because of the modification of tryptophan. We have found it convenient to either use amino acid analysis after acid hydrolysis or the ninhydrin reaction[35] after alkaline hydrolysis.[36]

The most frequent use of 2-hydroxy-5-nitrobenzyl bromide has been in the specific modification of tryptophan in peptides and proteins. Under appropriate reaction conditions (pH 4.0 or below), the reagent is highly specific for reaction with tryptophan (Figure 24). We have, on occasion, seen the modification of methionine residues under these conditions. This reagent also has the advantage of being a ''reporter'' group in the sense that the spectrum of the hydroxynitrobenzyl derivative is sensitive to changes in the microenvironment as shown in Figure 25. This decrease observed in absorbance at 410 nm associated with an increase in absorbance at 320 nm upon the addition of dioxane is similar to that seen with acidification and reflects the increase in the pKa of the phenolic hydroxyl group. The spectrophotometric titration curve for 2-hydroxy-5-nitrobenzyl alcohol is shown in Figure 26. Titration curves of oxidized and reduced laccase[37] which had been modified with 2-hydroxy-5-nitrobenzyl bromide are shown in Figure 27. From this experiment it was concluded that the tryptophanyl residues in laccase modified with 2-hydroxy-5-nitrobenzyl bromide are in an essentially aqueous microenvironment. The chemistry of the reaction of 2-hydroxy-5-nitrobenzyl bromide with tryptophan has been studied in some detail.[38] Disubstitution on the indole ring is a possibility and is usually seen as a sudden ''break'' in the plot of extent of modification vs. reagent excess (see Figure 28).

In our hands, the following procedure has been found useful. The protein or peptide to be modified is taken into 0.1 to 0.2 M sodium acetate buffer, pH 4 to 5. Reaction with other nucleophilic centers on the protein will become more of a problem as one approaches neutral pH. A 100-fold molar excess of 2-hydroxy-5-nitrobenzyl bromide (dissolved in a suitable water-miscible organic solvent such as acetone or dimethyl sulfoxide) is added in the dark. After 5 min (this time period was arbitrarily selected; the reaction can be considered to be essentially instantaneous to either modify tryptophan or undergo hydrolysis) the reaction mixture is taken by gel filtration into a solvent suitable for subsequent analysis. The extent of modification is determined under basic conditions as described above for the use of this reagent in the quantitative determination of tryptophan.

The use of 2-hydroxy-5-nitrobenzyl bromide does present problems in that the reagent is

---

*   Despite its wide use as a solvent for peptides and proteins, the use of formic acid is not recommended because of the potential of side reactions at amino functional groups; see Shively, J. E., Hawke, D., and Jones, B. N., Microsequence analysis of peptides and protein. III. Artifacts and effects of impurities on analysis, *Analyt. Biochem.*, 120, 312, 1982.

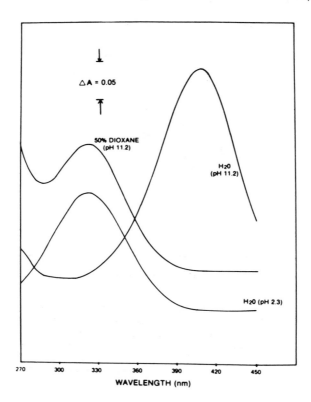

FIGURE 25. The UV absorption spectra of 2-hydroxy-5-nitro-benzyl alcohol (HNB-OH) in different solvents. The concentration of HNB-OH was 33.2 μM. (From Clemmer, J. D., Carr, J., Knaff, D. B., and Holwerda, R. A., *FEBS Lett.*, 91, 346, 1978. With permission.)

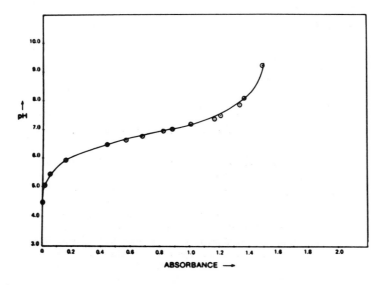

FIGURE 26. Spectrophotometric titration curve for 2-hydroxy-5-nitrobenzyl alcohol (HNB-OH). Shown is the absorbance of HNB-OH (76.4 μM) at 410 nm as a function of pH (phosphate buffers). (From Clemmer, J. D., Carr, J., Knaff, D. B., and Holwerda, R. A., *FEBS Lett.*, 91, 346, 1978. With permission.)

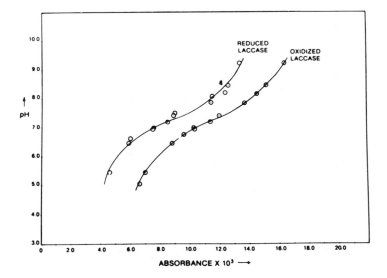

FIGURE 27.   Effect of pH on the absorbance (410 nm) of two forms of laccase modified with 2-hydroxy-5-nitrobenzyl bromide (HNB-laccase). Shown are spectrophotometric titration curves for oxidized and reduced HNB-laccase (2.5 μM; 0.38 mol HNB/mole laccase). The absorbance at 410 nm is shown as function of pH (phosphate buffers). (From Clemmer, J. D., Carr, J., Knaff, D. B., and Holwerda, R. A., *FEBS Lett.*, 91, 346, 1978. With permission.)

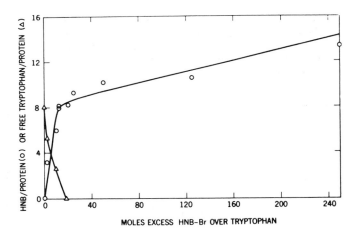

FIGURE 28.   The titration of the tryptophanyl residues of carboxymethyl chymotrypsinogen with 2-hydroxy-5-nitrobenzyl bromide. Carboxymethyl chymotrypsinogen (reduced and *S*-carboxymethylated with iodoacetic acid) was incubated in 10 *M* urea, pH 2.7, for 16 to 18 hr at which point 1 mℓ portions (5 mg of protein) were reacted with increasing amounts of 2-hydroxy-5-nitrobenzyl bromide dissolved in 0.1 mℓ acetone. The modified protein was separated from excess reagent by gel filtration (G-25 Sephadex) and subsequently analyzed for tryptophan (amino acid analysis after alkaline hydrolysis) and for the incorporation of the 2-hydroxy-5-nitrobenzyl group. (From Barman, T. E. and Koshland, D. E., Jr., *J. Biol. Chem.*, 242, 5771, 1967. With permission.)

$$R - \overset{\overset{\displaystyle O}{\|}}{C} - OR' + H_2O \longrightarrow R - \overset{\overset{\displaystyle O}{\|}}{C} - OH + HOR'$$

FIGURE 29.   The hydrolysis of an ester catalyzed by a serine protease.

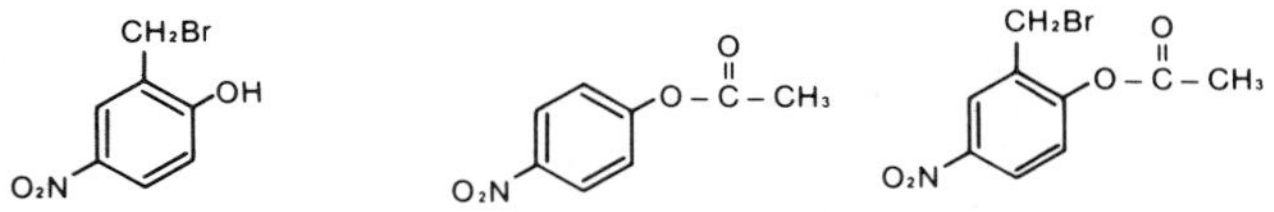

FIGURE 30.   The structures of 2-hydroxy-5-nitrobenzyl bromide (left), *p*-nitrophenyl acetate (center), and 2-acetoxy-5-nitrobenzyl bromide.

extremely sensitive to hydrolysis and is not very soluble under aqueous conditions. These difficulties are avoided and the characteristics of the reaction preserved by the use of the dimethyl sulfonium salt obtained from the reaction of 2-hydroxy-5-nitrobenzyl bromide with dimethyl sulfide.[39] This compound is easily synthesized or can be obtained from various commercial sources.

Horton and Koshland[40] have also developed a clever approach for modification of hydrolytic enzymes such as the serine proteases which catalyze the reaction shown in Figure 29. If 2-hydroxy-5-nitrobenzyl bromide is substituted at the phenolic hydroxyl, it is essentially unreactive as originally shown for the methoxy derivative. Horton and Young[41] prepared 2-acetoxy-5-nitrobenzyl bromide. This derivative, like the methoxy derivative, is essentially unreactive. There is considerable structural identity between 2-acetoxy-5-nitrobenzyl bromide and *p*-nitrophenyl acetate, which is a nonspecific substrate for chymotrypsin (Figure 30). α-Chymotrypsin removes the acetyl group from 2-acetoxy-5-nitrobenzyl bromide, thus generating 2-hydroxy-5-nitrobenzyl bromide at the active site which then either rapidly reacts with a neighboring nucleophile or undergoes hydrolysis. Uhteg and Lundblad[42] have used both the acetoxy and butyroxy derivatives in the study of thrombin. A similar approach has been used in the study of papain with 2-chloromethyl-4-nitrophenyl *N*-carbobenzoxy-glycinate.[43] It has been subsequently shown that this modification occurs at a specific tryptophan residue in papain.[44]

2-Hydroxy-5-nitrobenzyl bromide has been proved to be of use in the study of the functional role of tryptophan in the enzymes shown in Table 2.

Reagents with reaction characteristics similar to 2-hydroxy-5-nitrobenzyl bromide are the *o*-nitrophenylsulfenyl derivatives.[45] The reaction product resulting from the sulfonylation of lysozyme[46] in *o*-nitrobenzenesulfenyl chloride (2-nitrophenylsulfenyl chloride) (40-fold molar excess) pH 3.5 (0.1 *M* sodium acetate) has spectral characteristics which can be used to determine the extent of reagent incorporation (at 365 nm $\epsilon = 4 \times 10^{-3} M^{-1} cm^{-1}$). These reagents show considerable specificity for the modification of tryptophan at pH ≤4.0 (Figure 31). Possible side reactions with other nucleophiles such as amino groups need to be considered. In the case of human chorionic somatomammotropin and human pituitary growth hormone,[47] reaction with *o*-nitrophenylsulfenyl chloride (2-nitrophenylsulfenyl chloride) was achieved in 50% acetic acid but not in 0.1 sodium acetate, pH 4.0. Wilchek and Miron[48] have reported on the reaction of 2,4-dinitrophenylsulfenyl chloride with tryptophan in peptides and protein and subsequent conversion of the modified tryptophan to 2-thiotryptophan by reaction with β-mercaptoethanol at pH 8.0 (see Figure 33). The thiolysis of the modified tryptophan is responsible for changes in the spectral properties of the derivative. The characteristics of the modified tryptophan have resulted in the development of a facile purification scheme for peptides containing the modified tryptophan residues.[49,50]

Table 2
EXAMPLES OF THE MODIFICATION OF PROTEINS WITH 2-HYDROXY-5-NITROBENZYL BROMIDE

Protein	Solvent	Molar excess	Residues modified	Ref.
Pepsin	0.1 *M* NaCl[a]	300	2/4	1
Streptococcal proteinase	0.46 *M* sodium phosphate, pH 3.1	200	1.8/4	2
Pancreatic deoxyribonuclease	0.050 *M* CaCl$_2$[b]	100	1/3	3
Carbonic anhydrase	0.1 *M* phosphate, pH 6.8	100	—[c]	4
Trypsin	0.1 *M* NaCl, 0.02 *M* CaCl$_2$[d] pH 4.2 (pH-Stat)	ca. 100	1/4	5
Human chorionic somatomammotropin	0.05 *M* glycine, pH 2.8	—	—	6
Naja naja neurotoxin	0.2 *M* acetic acid[e]	40	—[f]	7
Glyceraldehyde-3-phosphate dehydrogenase	pH 6.75[g]	30[h]	1/3	8
α-Mannosidase (*Phaseolus vulgaris*)	0.1 *M* sodium acetate, pH 3.7	100	5/28	9
Thrombin	0.2 *M* acetate, pH 4.0	100	1/8	10
Laccase	pH 6.95[i]	50	0.30/6	11
	pH 4.00[i]	50	0.58/6	11
	pH 3.30[i]	110	2.39/6	11
Human serum albumin	10 *M* urea, pH 4.4	1000	1.1/1[j]	12

[a]   pH Adjusted with 50% acetic acid.
[b]   pH Remained between 4.0 an 4.5 without need for buffer.
[c]   Variation with respect to enzyme source.
[d]   pH Maintained at 4.2 by addition of NaOH (pH Stat).
[e]   pH 2.7.
[f]   Polymerization occurred.
[g]   pH Maintained at 6.75 by the addition of 0.1 *M* NaOH.
[h]   Dimethyl (2-hydroxy-5-nitrobenzyl) sulfonium bromide was used in the experiments. Prior to reaction, the active site sulfhydryl was blocked by reaction with 5,5'-dithiobis(2-nitrobenzoate).
[i]   Unbuffered, pH maintained by titration with NaOH.
[j]   Incorporation determined at pH 7.4 after the following relationship:[55] moles 2-hydroxy-5-nitrobenzyl bromide per mole albumin = $(A_{410} \times 69,000 \times 0.498)/13,800 \times (A_{280} - 0.167) \times A_{410}$.

References for Table 2

1. **Dopheide, T. A. A. and Jones, W. M.,** Studies on the tryptophan residues in porcine pepsin, *J. Biol. Chem.*, 243, 3906, 1968.
2. **Robinson, G. W.,** Reaction of a specific tryptophan residue in streptococcal proteinase with 2-hydroxy-5-nitrobenzyl bromide, *J. Biol. Chem.*, 245, 4832, 1970.
3. **Poulos, T. L. and Price, P. A.,** The identification of a tryptophan residue essential to the catalytic activity of bovine pancreatic deoxyribonuclease, *J. Biol. Chem.*, 246, 4041, 1971.
4. **Lindskog, S. and Nilsson, A.,** The location of tryptophanyl groups in human and bovine carbonic anhydrases. Ultraviolet difference spectra and chemical modification, *Biochim. Biophys. Acta*, 295, 117, 1973.
5. **Imhoff, J. M., Keil-Dlouha, V., and Keil, B.,** Functional changes in bovine α- and β-trypsins caused by the substitution of tryptophan-199, *Biochimie*, 55, 521, 1973.
6. **Neri, P., Arezzini, C., Botti, R., Cocola, F., and Tarli, P.,** Modification of the tryptophanyl residue and its effect on the immunological and biological activity of human chorionic somatomammotropin, *Biochim. Biophys. Acta*, 322, 88, 1973.
7. **Karlsson, E., Eaker, D., and Drevin, H.,** Modification of the invariant tryptophan residue of two *Naja naja* neurotoxins, *Biochim. Biophys. Acta*, 328, 510, 1973.
8. **Heilman, H. D. and Pfleiderer, G.,** On the role of tryptophan residues in the mechanism of action of glyceraldehyde-3-phosphate dehydrogenase as tested by specific modification, *Biochim. Biophys. Acta*, 384, 331, 1975.

Table 2 (continued)

9. **Paus, E.,** The chemical modification of tryptophan residues of α-mannosidase from *Phaseolus vulgaris,* *Biochim. Biophys. Acta,* 533, 446, 1978.

10. **Uhteg, L. C.and Lundblad, R. L.,** The modification of tryptophan in bovine thrombin, *Biochem. Biophys. Acta,* 491, 551, 1977.

11. **Clemmer, J. D., Carr, J., Knaff, D. B., and Holwerda, R. A.,** Modification of laccase tryptophan residues with 2-hydroxy-5-nitrobenzyl bromide, *FEBS Lett.,* 91, 346, 1978.

12. **Fehske, K. J., Müller, W. E., and Wollert, U.,** The modification of the lone tryptophan residue in human serum albumin by 2-hydroxy-5-nitrobenzyl bromide. Characterization of the modified protein and the binding of L-tryptophan and benzodiazepines to the tryptophan-modifed albumin, *Hoppe Seyler's Z. Physiol. Chem.,* 359, 709, 1978.

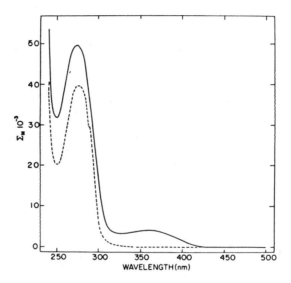

FIGURE 31. The UV absorbance spectrum of egg-white lysozyme after modification of tryptophan-62 with 2-nitro-phenylsulfenyl chloride (*o*-nitrobenzenesulfenyl chloride). Shown is the UV absorption spectrum of 1-NPS-lysozyme-(---), and native lysozyme (---). The measurements were performed in water at pH 7.0.

A heterobifunctional reagent based on the *o*-nitrophenylsulfenyl compounds has been recently developed by Demoliou and Epand.[51] These investigators reported the synthesis of 2-nitro-4-azidophenylsulfenyl chloride and the subsequent use of this reagent for preparation of a photoreactive glucagon derivative.

An example of a particularly good study on the modification of tryptophan in proteins and peptides is that of Hazum and co-workers on luteinizing–hormone–releasing hormone.[52] By the use of a number of the procedures described above, these investigators were able to conclude that alterations in the indole ring which decrease the electron density at position 3 are associated with severe loss in biological activity.

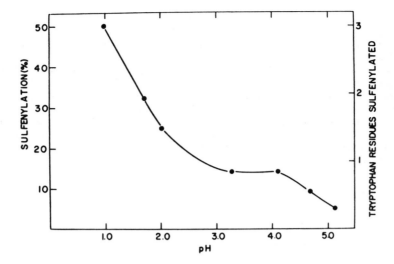

FIGURE 32.   The extent of sulfenylation of tryptophanyl residues in egg-white lysozyme by 2-nitrophenylsulfenyl chloride as function of the pH. Sulfenylation was carried out at protein concentration of 0.5 μmol in 1 mℓ of 0.1 *M* buffered solutions (HCl-KCl at pH 1 to 2; sodium acetate at pH 3 to 5) with 20 μmol of 2-nitrophenylsulfenyl chloride for 5 hr. (From Shechter, Y., Burstein, Y., and Patchornik, A., *Biochemistry,* 11, 653, 1972. With permission.)

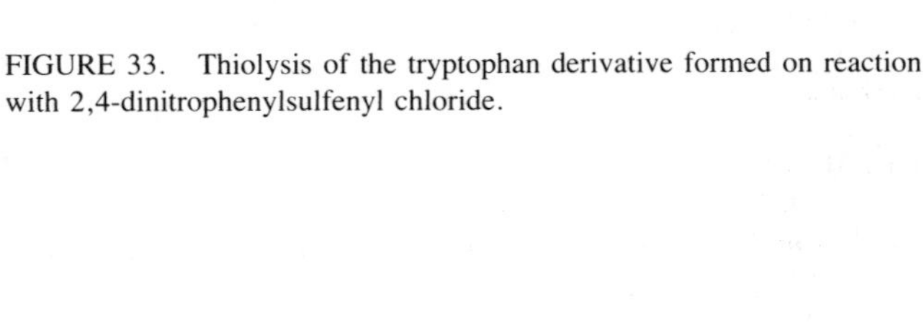

FIGURE 33.   Thiolysis of the tryptophan derivative formed on reaction with 2,4-dinitrophenylsulfenyl chloride.

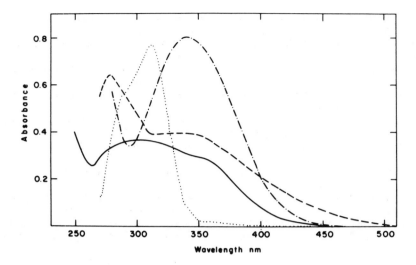

FIGURE 34. UV absorption spectra of the derivatives formed by the thiolysis of 2-(2,4-dinitrophenylsulfenyl) tryptophan. Shown are the UV absorption spectra of 2,4-dinitrophenylsulfenyltryptophan (----), 2-thioltryptophan (····), ditryptophanyl-2-disulfide (——), and S-2,4-dinitrophenyl-2-mercaptoethanol (—·—). The spectral studies were performed in 0.1 *M* ammmonium bicarbonate. (From Wilchek, M. and Miron, T., *Biochem. Biophys. Res. Commun.*, 47, 1015, 1972. With permission.)

REFERENCES

1. **Liu, T.-Y. and Chang, Y. H.,** Hydrolysis of proteins with *p*-toluenesulfonic acid. Determination of tryptophan, *J. Biol. Chem.,* 246, 2842, 1971.
2. **Simpson, R. J., Neuberger, M. R., and Liu, T.-Y.,** Complete amino acid analysis of proteins from a single hydrolysate, *J. Biol. Chem.,* 251, 1936, 1976.
3. **Hachimori, Y., Horinishi, H., Kurihara, K., and Shibata,K.,** States of amino residues in proteins. V. Different reactivities with H_2O_2 of tryptophan residues in lysozyme, proteinases and zymogens, *Biochim. Biophys. Acta,* 93, 346, 1964.
4. **Kotoku, I., Matsushima, A., Bando, M., and Inada, Y.,** Tyrosine and tryptophan residues and amino groups in thrombin related to enzymic activities, *Biochim. Biophys. Acta,* 214, 490, 1970.
5. **Sanda, A. and Irie, M.,** Chemical modification of tryptophan residues in ribonuclease from a *Rhizopus* sp., *J. Biochem.,* 87, 1079, 1980.
6. **Matsushima, A., Takiuchi, H., Saito, Y., and Inada, Y.,** Significance of tryptophan residues in the D-domain of the fibrin molecule in fibrin polymer formation, *Biochim. Biophys. Acta,* 625, 230, 1980.
7. **Patchornik, A., Lawson, W. B., and Witkop, B.,** Selective cleavage of peptide bonds. I. Mechanism of oxidation of β-substituted indoles with *N*-bromosuccinimide, *J. Am. Chem. Soc.,* 80, 4747, 1958.
8. **Spande, T. F. and Witkop, B.,** Determination of the tryptophan content of protein with *N*-bromosuccinimide, *Meth. Enzymol.,* 11, 498, 1967.
9. **Spande, T. F. and Witkop, B.,** Tryptophan involvement in the function of enzymes and protein hormones as determined by selective oxidation with *N*-bromosuccinimide, *Meth. Enzymol.,* 11, 506, 1967.
10. **Spande, T. F. and Witkop, B.,** Tryptophan involvement in binding sites of proteins and in enzyme–inhibitor complexes as determined by oxidation with *N*-bromosuccinimide, *Meth. Enzymol.,* 11, 522, 1967.
11. **Green, N. M.,** Avidin. 3. The nature of the biotin binding site, *Biochem. J.,* 89, 599, 1963.
12. **Crestfield, A. M., Moore, S., and Stein, W. H.,** The preparation and enzymatic hydrolysis of reduced and S-carboxymethylated proteins, *J. Biol. Chem.,* 238, 622, 1963.
13. **Ohnishi, M., Kawagishi, T., Abe, T., and Hiromi, K.,** Stopped-flow studies on the chemical modification with *N*-bromosuccinimide of model compounds of tryptophan residues, *J. Biochem.,* 87, 273, 1980.

14. **Sartin, J. L., Hugli, T. E., and Liao, T.-H.,** Reactivity of the tryptophan residues in bovine pancreatic deoxyribonuclease with *N*-bromosuccinimide, *J. Biol. Chem.,* 255, 8633, 1980.

15. **Daniel, V. W., III and Trowbridge, C. G.,** The effect of *N*-bromosuccinimide upon trypsinogen activation and trypsin catalysis, *Arch. Biochem. Biophys.,* 134, 506, 1969.

16. **Freisheim, J. H. and Huennekens, F. M.,** Effect of *N*-bromosuccinimide on dihydrofolate reductase, *Biochemistry,* 8, 2271, 1969.

17. **Warwick, P. E., D'Souza, L., and Freisheim, J. H.,** Role of tryptophan in dihydrofolate reductase, *Biochemistry,* 11, 3775, 1972.

18. **Poulos, T. L. and Price, P. A.,** The identification of a tryptophan residue essential to the catalytic activity of bovine pancreatic deoxyribonuclease, *J. Biol. Chem.,* 246, 4041, 1971.

19. **Schwabe, C. and Braddon, S. A.,** Evidence for the essential tryptophan residue at the active site of relaxin, *Biochem. Biophys. Res. Commun.,* 68, 1126, 1976.

19a. **Nagami, K.,** The participation of a tryptophan residue in the binding of ferric iron to pyrocatechase, *Biochem. Biophys. Res. Commun.,* 51, 364, 1973.

20. **Kosman, D. J., Ettinger, M. J., Bereman, R. D., and Giordano, R. S.,** Role of tryptophan in the spectral and catalytic properties of the copper enzyme, galactose oxidase, *Biochemistry,* 16, 1597, 1977.

21. **Fujimori, H., Ohnishi, M., and Hiromi, K.,** Tryptophan residues of saccharifying α-amylase from *Bacillus subtilis*. A kinetic discrimination of states of tryptophan residues using *N*-bromosuccinimide, *J. Biochem.,* 83, 1503, 1978.

22. **Inokuchi, N., Takahashi, T., Yoshimoto, A., and Irie, M.,** *N*-Bromosuccinimide oxidation of a glucoamylase from *Aspergillus saitoi, J. Biochem.,* 91, 1661, 1982.

23. **O'Gorman, R. B. and Matthews, K. S.,** *N*-Bromosuccinimide modification of lac repressor protein, *J. Biol. Chem.,* 252, 3565, 1977.

24. **Kördel, W. and Schneider, F.,** Chemical modification of two tryptophan residues abolishes the catalytic activity of aminoacylase, *Hoppe-Seyler's Z. Physiol. Chem.,* 357, 1109, 1976.

25. **Ramachandran, L. K. and Witkop, B.,** *N*-Bromosuccinimide cleavage of peptides, *Meth. Enzymol.,* 11, 283, 1967.

26. **Feldhoff, R. C. and Peters, T., Jr.,** Determination of the number and relative position of tryptophan residues in various albumins, *Biochem. J.,* 159, 529, 1976.

27. **Shechter, Y., Patchornik, A., and Burstein, Y.,** Selective chemical cleavage of tryptophanyl peptide bonds by oxidative chlorination with *N*-chlorosuccinimide, *Biochemistry,* 15, 5071, 1976.

28. **Coletti-Previero, M.-A., Previero, A., and Zuckerkandl, E.,** Separation of the proteolytic and esteratic activities of trypsin by reversible structural modifications, *J. Mol. Biol.,* 39, 493, 1969.

29. **Previero, A., Coletti-Previero, M.-A., and Cavadore, J. C.,** A reversible chemical modification of the tryptophan residue, *Biochim. Biophys. Acta,* 147, 453, 1967.

30. **Holmgren, A.,** Reversible chemical modification of the tryptophan residues of thioredoxin from *Eschericia coli* B., *Eur. J. Biochem.,* 26, 528, 1972.

31. **Previero, A., Prota, G., and Coletti-Previero, M.-A.,** C-Acylation of the tryptophan indole ring and its usefulness in protein chemistry, *Biochim. Biophys. Acta,* 285, 269, 1972.

32. **Koshland, D. E., Jr., Karkhanis, Y. D., and Latham, H. G.,** An environmentally-sensitive reagent with selectivity for the tryptophan residue in proteins, *J. Am. Chem. Soc.,* 86, 1448, 1964.

33. **Horton, H. R. and Koshland, D. E., Jr.,** A highly reactive colored reagent with selectivity for the tryptophanyl residue in proteins, 2-Hydroxy-5-nitrobenzyl bromide, *J. Am. Chem. Soc.,* 87, 1126, 1965.

34. **Barman, T. E. and Koshland, D. E., Jr.,** A colorimetric procedure for the quantitative determination of tryptophan residues in proteins, *J. Biol. Chem.,* 242, 5771, 1967.

35. **Moore, S.,** Amino acid analysis: aqueous dimethyl sulfoxide as solvent for the ninhydrin reaction, *J. Biol. Chem.,* 243, 6281, 1968.

36. **Fruchter, R. G. and Crestfield, A. M.,** Preparation and properties of two active forms of ribonuclease dimer, *J. Biol. Chem.,* 240, 3868, 1965.

37. **Clemmer, J. D., Carr, J., Knaff, D. B., and Holwerda, R. A.,** Modification of laccase tryptophan residues with 2-hydroxy-5-nitrobenzyl bromide, *FEBS Lett.,* 91, 346, 1978.

38. **Loudon, G. M. and Koshland, D. E., Jr.,** The chemistry of a reporter group: 2-hydroxy-5-nitrobenzyl bromide, *J. Biol. Chem.,* 245, 2247, 1970.

39. **Horton, H. R. and Tucker, W. P.,** Dimethyl (2-hydroxy-5-nitrobenzyl) sulfonium salts. Water-soluble environmentally sensitive protein reagents, *J. Biol. Chem.,* 245, 3397, 1970.

40. **Horton, H. R. and Koshland, D. E., Jr.,** Reactions with reactive alkyl halides, *Meth. Enzymol.,* 11, 556, 1967.

41. **Horton, H. R. and Young, G.,** 2-Acetoxy-5-nitrobenzyl chloride. A reagent designed to introduce a reporter group near the active site of chymotrypsin, *Biochim. Biophys. Acta,* 194, 272, 1969.

42. **Uhteg, L. C. and Lundblad, R. L.,** The modification of tryptophan in bovine thrombin, *Biochim. Biophys. Acta,* 491, 551, 1977.

43. **Mole, J. E. and Horton, H. R.,** A kinetic analysis of the enhanced catalytic efficiency of papain modified by 2-hydroxy-5-nitrobenzylation, *Biochemistry,* 12, 5285, 1973.
44. **Chang, S.-M. T. and Horton, H. R.,** Structure of papain modified by reaction with 2-chloromethyl-4-nitrophenyl *N*-carbobenzoxylglycinate, *Biochemistry,* 18, 1559, 1979.
45. **Fontana, A. and Scoffone, E.,** Sulfenyl halides as modifying reagents for polypeptides and proteins, *Meth Enzymol.,* 25, 482, 1972.
46. **Shechter, Y., Burstein, Y., and Patchornik, A.,** Sulfenylation of tryptophan-62 in hen egg-white lysozyme, *Biochemistry,* 11, 653, 1972.
47. **Bewley, T. A., Kawauchi, H., and Li, C. H.,** Comparative studies of the single tryptophan residue in human chorionic somatomammotropin and human pituitary growth hormone, *Biochemistry,* 11, 4179, 1972.
48. **Wilchek, M. and Miron, T.,** The conversion of tryptophan to 2-thioltryptophan in peptides and proteins, *Biochem. Biophys. Res. Commun.,* 47, 1015, 1972.
49. **Chersi, A. and Zito, R.,** Isolation of tryptophan-containing peptides by adsorption chromatography, *Anal. Biochem.,* 73, 471, 1976.
50. **Rubinstein, M., Schechter, Y., and Patchornik, A.,** Covalent chromatography — the isolation of tryptophanyl containing peptides by novel polymeric reagents, *Biochem. Biophys. Res. Commun.,* 70, 1257, 1976.
51. **Demoliou, C. D. and Epand, R. M.,** Synthesis and characterization of a heterobifunctional photoaffinity reagent for modification of tryptophan residues and its application to the preparation of a photoreactive glycagon derivative, *Biochemistry,* 19, 4539, 1980.
52. **Hazum, E., Fridkin, M., Meidan, R., and Koch, Y.,** On the role of tryptophan in luteinizing-hormone-releasing hormone (luliberin), *Eur. J. Biochem.,* 79, 269, 1977.

Chapter 3

THE MODIFICATION OF TYROSINE

The specific modification of tyrosyl residues (Figure 1) in proteins has provided considerable information regarding the participation of these residues in the catalytic processes of enzymes as well as specific binding processes of proteins. There has also been considerable interest in the modification of tyrosyl residues to introduce spectral probes into proteins such as the modification of tyrosyl residues with aromatic diazonium compounds or tetranitromethane.

There are a number of reagents which may result in the modification of tyrosyl residues. The two most frequently used modifications of tyrosyl residues in proteins are reaction with *N*-acetylimidizole to form the *O*-acetyl derivative and nitration with tetranitromethane to form the 3-nitro derivative. Each of these modifications is considered in some detail later in this chapter.

There is considerable literature concerning the reaction of tyrosyl residues with aromatic diazonium compounds.[1-3] Diazonium salts readily couple with proteins to form colored derivatives with interesting spectral properties. Reaction with diazonium salts is accomplished at alkaline pH (pH 8 to 9, bicarbonate/carbonate or borate buffers). It is relatively difficult to obtain specific residue class modification with the aromatic diazonium salts but tyrosine, lysine, and histidine are rapidly modified.[4,5] Largely as a result of this lack of specificity, the use of this class of reagents has been somewhat limited. The reaction of tyrosyl residues with diazotized arsanilic acid is shown in Figure 2.

The reaction of chymotrypsinogen A with diazotized arsanilic acid has been investigated.[6] Diazotization of arsanilic acid is accomplished by treatment of *p*-arsanilic acid with nitrous acid (0.55 mM sodium nitrite in 0.15 M HCl at 0°C). After adjustment of the pH to 5.5 with NaOH the reagent is diluted to a final concentration of 0.02 M. Reaction with chymotrypsinogen is accomplished in 0.5 M sodium bicarbonate buffer, pH 8.5 with a 20-fold excess of reagent at 0°C. The reaction is terminated by the addition of a sufficient quantity of aqueous phenol (0.1 M) to react with excess reagent. The extent of the formation of monoazotyrosyl and monoazohistidyl derivatives is determined by spectral analysis.[4,5] The extent of reagent incorporation is determined by atomic absorption analysis for arsenic. Tyrosine ($\sim$ 1.0 mol/mol) and lysine ($\sim$ 4 mol/mol) were the only amino acid residues modified to any significant extent under these reaction conditions. The arsaniloazo functional group provides a spectral probe that can be used to study conformational change in proteins. In this particular study, there was a substantial change in the circular dichroism spectrum during the activation of the modified chymotrypsinogen preparation by trypsin. It is of interest that the modification of chymotrypsinogen by diazotized arsanilic acid does not apparently affect either the rate of activation or amount of potential catalytic activity as judged by the hydrolysis of *N*-benzoyl-L-tyrosine ethyl ester.

The reaction of α-chymotrypsin with three diazonium salt derivatives (analogues) of *N*-acetyl-D-phenylalanine methyl ester[7] has been studied. The corresponding aromatic amine was converted to the diazonium salt by the action of nitrous acid (sodium nitrite per 0.6 M HCl at 0°C) and, after neutralization (NaOH) and dilution with 0.2 M sodium borate, pH 8.4, was used immediately for the modification of α-chymotrypsin (diazonium salt at a tenfold molar excess) in 0.2 M sodium borate, pH 8.4 at ambient temperature for 1 hr. The reaction was terminated by gel filtration (G-25 Sephadex) in 0.001 M HCl. Amino acid analysis showed that only tyrosine is modified under these reaction conditions. Subsequent analysis showed that Tyr^{146} is modified by each of the three reagents. It is of interest to note the peptide with the modified tyrosine residue (possessing a yellow color) absorbs to the gel filtration matrix (G-10 equilibrated with 0.001 M HCl) and is eluted with 50% acetic acid.

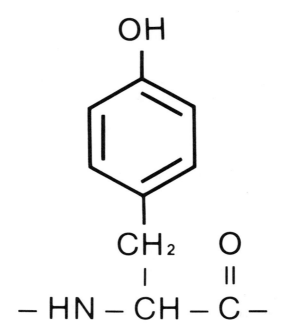

FIGURE 1.   The structure of tyrosine.

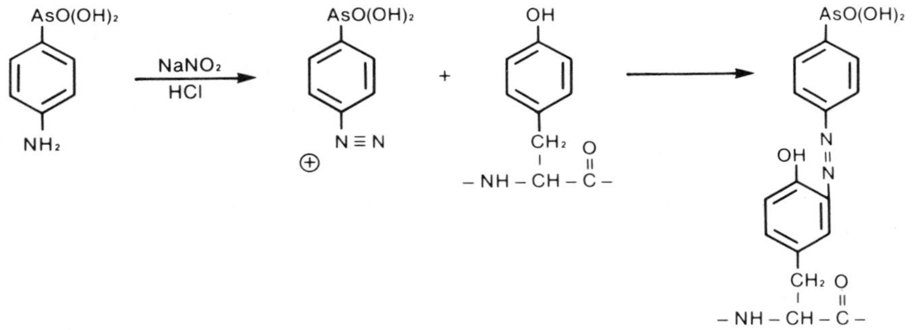

FIGURE 2.   The formation of diazotized arsanilic acid and the modification of tyrosine with this reagent.

(This phenomenon is somewhat similar to that observed with tryptophan-containing peptides which have been modified with 2-hydroxy-5-nitrobenzyl bromide.)[8] Modification of chymotrypsin with these reagents had varying effects on chymotryptic activity.

Pancreatic ribonuclease has been modified by a diazonium salt derivative of uridine $2'$ $(3')5'$-diphosphate.[9] Modification occurs at a specific tyrosine residue (Tyr[73]). Modification of ribonuclease with $5'$-(4-diazophenyl phosphoryl)–uridine $2'(3')$-phosphate was accomplished by *in situ* generation of the diazonium salt from the corresponding amine by NaNO$_2$/HCl in the cold. The pH was then adjusted to pH 8.4 (NaOH); the solution was added to ribonuclease in 0.1 M borate, pH 8.4 and the reaction allowed to proceed for 1 hr at ambient temperature. The reaction was terminated by gel filtration (G-25 Sephadex) in 0.1 M acetic acid. The extent of modification was determined by spectral analysis and by amino acid analysis. Tyrosine was the only amino acid residue modified. Although it is relatively easy

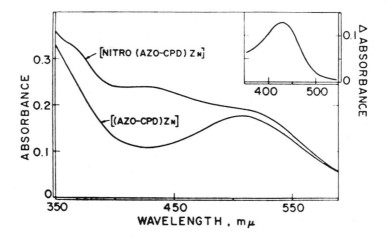

FIGURE 3.   The UV absorption spectra of azocarboxypeptidase and nitroazo-carboxypeptidase. Azocarboxypeptidase was obtained by the reaction of carboxypeptidase A with a sevenfold molar excess of diazonium-1H-tetrazole in 1.0 M NaCl — 0.067 M potassium bicarbonate/carbonate, pH 8.8, at 0 to 4°C; the reaction was quenched after 30 min by the addition excess Tris-Cl, pH 8.0 and excess reagent removed by dialysis. Nitroazocarboxypeptidase was obtained by the reaction of tetranitromethane with carboxypeptidase previously modified with a sevenfold molar excess of diazonium-1H-tetrazole as described above. Shown is the absorption spectra of azocarboxypeptidase ((AZO CPD)ZN) and nitroazo-carboxypeptidase (NITRO (AZO CPN)ZN) in 0.1 N NaOH both at 100 μM. The inset represents the difference spectrum of nitroazocarboxypeptidase minus azo-carboxypeptidase. (From Riordan, J. F., Sokolovsky, M., and Vallee, B. L., *Biochemistry*, 6, 3609, 1967. With permission.)

to assess the loss of tyrosyl residues, precise determination of diazotization can be obtained only after reduction to the corresponding amine with sodium sulfite. These investigators also examined the reaction of ribonuclease with *p*-diazophenylphosphate under the same conditions of solvent and temperature. Reaction with this reagent was far less specific, with losses of lysine, histidine, and tyrosine (3 mol/mol ribonuclease).

Reaction of bovine carboxypeptidase A with various diazonium salts has been explored in greater detail than that of the above proteins. Vallee and co-workers[10,11] reported on the reaction of bovine carboxypeptidase A crystals with diazotized *p*-arsanilic acid (conditions not specified) and obtained specific modification of Tyr[248]. Purification of the peptide containing the modified tyrosine residue was achieved by using antibody directed against the arsaniloazotyrosyl group. The antibodies were obtained from rabbits using arsaniloazoval-bumin and arsaniloazobovine γ-globulin as antigen. The reaction of bovine carboxypeptidase A with diazotized 5-amino-1H-tetrazole has been reported.[12] Diazotized 5-amino-1H-tetrazole also specifically reacts with Tyr[248] in bovine carboxypeptidase A (in 0.67 M potassium bicarbonate/carbonate, 1.0 M NaCl, pH 8.8). A sevenfold molar excess of reagent was used and the reaction terminated after 30 min by the addition of Tris buffer. The extent of modification of tyrosine to tetrazolylazotyrosine is determined by absorbance at 483 nm (Figure 3) ($\epsilon = 8.7 \times 10^3\ M^{-1}\ cm^{-1}$). Modification of Tyr[248] in carboxypeptidase A by this reagent permits the subsequent modification of Tyr[198] by tetranitromethane.

Iodination is somewhat infrequently used for the modification of tyrosyl residues in protein. The reaction is still of considerable value since the process of the radiolabeling of proteins with either of the iodine radioisotopes (^{125}I, ^{131}I) primarily involves the modification of tyrosine residues in proteins. It is, of course, of critical importance to appreciate the strength of the elemental halides as oxidizing agents.

Nevertheless, iodination via various vehicles has proved to be a useful approach to the modification of tyrosyl residues.[13] Filmer and Koshland[14] explored the iodination of "sulfoxide-chymotrypsin", a derivative in which methionine has been selectively oxidized to the sulfoxide. Under reaction conditions utilized by these investigations, 0.2 M glycine, pH 8.5 at 0°C tyrosine was the most sensitive amino acid residue with modification at methionine, histidine, and tryptophan occurring only after tyrosine modification was 95% + complete. Another group of investigators also explored the iodination of tyrosyl residues in chymotrypsin[15] using reaction conditions primarily developed in their laboratory,[16] pH 8.0 at 0°C, pH maintained with Tris buffer (0.67 M). Iodination has been utilized to study the reactivity of tyrosyl residues in cytochrome b_5.[17] Iodination is accomplished with a tenfold molar excess of I_2 (15 mM I_2 in 30 mM KI) in 0.025 M sodium borate, pH 9.8. Iodination with limiting amounts of iodine is accomplished with a two- to six-fold molar excess of iodine in 0.020 M potassium phosphate, pH 7.5 at 0°C. Monoiodination and diiodination of tyrosyl residues is observed. Iodination with a tenfold molar excess of I_2 results in the formation of 3 mol of diiodotyrosine per mole of cytochrome 6_5. The fourth tyrosyl residue is modified only in the presence of 4.0 M urea. Iodination of tyrosine results in a decrease in the pKa of the phenolic hydroxyl groups.[18] Iodination with a limiting amount of iodine as described above results first in the formation of 2 mol of monoiodotyrosine, and then 1 mol of diiodotyrosine, and 1 mol of monoiodotyrosine. Tyrosyl residues which can be iodinated are also available for 0-acetylation with acetic anhydride (0.1 M potassium phosphate, pH 7.5; acetic anhydride added in two portions over 1 hr at 0°C; maintained at pH 7.8 with NaOH (1 M) addition).

The modification of tyrosyl residues in phosphoglucomutase by iodination has been reported.[19] Modification is achieved by reaction in 0.1 M borate, pH 9.5 with 1 mM I_2 (obtained by an appropriate dilution of a stock iodine/iodide solution, 0.05 M I_2 in 0.24 M KI) at 0°C for 10 min (protein concentration of 2 × $10^{-5}M$).[20] Complete loss of enzymatic activity was observed with these reaction conditions, but the stoichiometry of modification was not established. Nitration of 7/20 tyrosyl residues resulted in 83% loss of catalytic activity. These investigators also studied the reaction of phosphoglucomutase with diazotized sulfanilic acid and N-acetylimidazole.

The above modifications utilize reaction of tyrosyl residues in proteins with iodine/iodide solutions at alkaline pH. Iodination of tyrosyl residues can also be accomplished with iodine monochloride (ICl) at mildly alkaline pH. One such study explores the modification of galactosyltransferase.[21] The modification is accomplished by reaction in 0.2 M sodium borate, pH 8.0. The reaction is initiated by desired amount of a stock solution of ICl.[22] A stock solution of 0.02 M ICl is prepared by adding 21 mℓ 11.8 M HCl (stock concentrated HCl) to approximately 150 mℓ of H_2O containing 0.555 g KCl, 0.3567 g KIO_3, and 29.23 g NaCl. The solvent is taken to a final volume of 250 mℓ with H_2O. Free iodine is then extracted with CCl_4, if necessary, and the solution aerated to remove trace amounts of CCl_4. The resulting solution of ICl is stable for an indefinite period of time under ambient conditions. Reaction proceeds for 1 min at ambient temperature and is terminated by the addition of a 1/6 volume of 0.5 M $Na_2S_2O_3$ (50 $\mu\ell$ for a 0.300 mℓ reaction mixture). Radiolabeled sodium iodide ($Na^{125}I$) is included to provide a mechanism for establishing the stoichiometry of the reaction. The reaction mixture, after the addition of $Na_2S_2O_3$, is subjected to gel filtration on Bio-Gel P-10 in 0.1 M Tris, pH 7.4. In experiments designed to assess the relationships between reagent (ICl) concentration and the extent of modification, a maximum of 10 g-atom of iodine were incorporated into galactosyltransferase at a 40-fold molar excess of reagent. Incorporation of iodine is linear up to this excess of reagent and slowly declines at higher concentration of ICl. Modification of tryptophanyl residues was excluded by direct analysis, and the only iodinated amino acids obtained from the modified protein were monoiodotyrosine and diiodotyrosine. Modification of other residues such as histidine and methionine by oxidation without incorporation of iodine was not excluded.

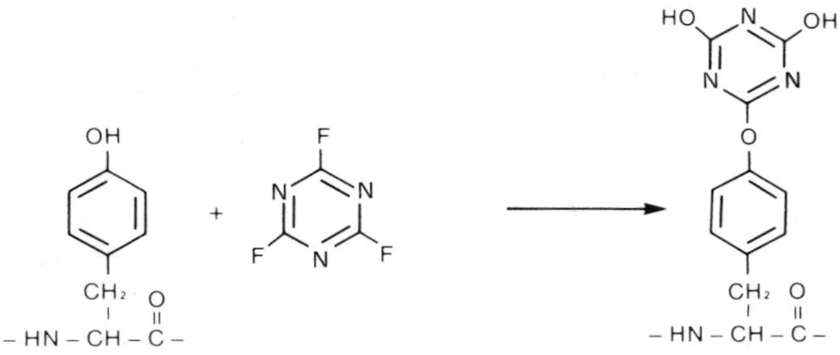

FIGURE 4. A scheme for the modification of tyrosyl residues with cyanuric fluoride.

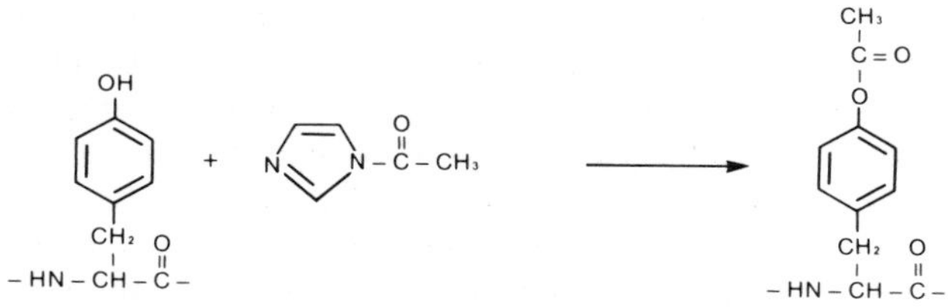

FIGURE 5. A scheme for the reaction of tyrosyl residues with *N*-acetylimidazole.

Iodination can also be accomplished by peroxidase per H_2O_2 per NaI. A recent procedure was described for the modification of tyrosyl residues in insulin.[23] In these studies 20 mg of porcine insulin in 20 mℓ 0.4 M sodium phosphate, 6.0 M urea, pH 7.8 was combined with 10 mℓ Na^{125}I (1 mCi) and 3.6 mg urea, and H_2O_2 (5 $\mu\ell$ 0.3 mM solution) and peroxidase (Sigma, 0.2 mg/mℓ; 5 $\mu\ell$) added. The preparative reaction was terminated by dilution with an equal volume of 40% (w/v) sucrose. It is of interest to note that the enzyme-catalyzed iodination proceeds with efficiency in 6.0 M urea. Iodination of tyrosyl residues in peptides and proteins can also be accomplished with chloramine T.[24,25]

Tyrosyl residues in proteins are also modified by reaction with cyanuric fluoride (Figure 4).[26,27] The reaction proceeds at alkaline pH (9.1) via modification of the phenolic hydroxyl group with a change in the spectral properties of tyrosine. The phenolic hydroxyl groups must be ionized (phenoxide ion) for reaction with cyanuric fluoride. The modification of tyrosyl residues in elastase[28] and yeast hexokinase[29] with cyanuric fluoride has been reported.

Modification of tyrosyl residues can occur as an interesting side reaction with other residue-specific reagents such as 7-chloro-4-nitrobenzo-2–oxa-1,3-diazole[30,31] and 1-fluoro-2,4-dinitrobenzene.[32] Of particular interest is the reaction of tyrosyl residues with diisopropyl-phosphofluoridate to form the *O*-diisopropylphosphoryl derivative.[33-36]

Although the above reagents have proved useful for the selective modification of tyrosyl residues in proteins, the majority of the studies in this area have utilized reaction with either *N*-acetylimidazole and/or tetranitromethane.

The development of *N*-acetylimidazole as a reagent (Figure 5) for the selective modification of tyrosyl residues can, in part, be traced to the early observations[37-39] that *N*-acetylimidazole is, in fact, an energy-rich compound. The preparation of *O*-acyl derivatives via the action of carboxylic acid anhydrides (i.e., acetic anhydride) has been used for some time, but it is very difficult to obtain selective modification of tyrosine as these reagents readily react with primary amines to form stable *N*-acyl derivatives.[40,41] It is, however, possible to obtain the selective modification of tyrosine with acetic anhydride by reaction at mildly acidic pH

(1.0 M acetate, pH 5.8, 25°C; approximate 20,000-fold molar excess of acetic anhydride (5.1 × 10^{-2} M acetic anhydride, 2.9 × 10^{-6} M enzyme).[42]

N-acetylimidazole was first used as a reagent for the modification of tyrosyl residues in bovine pancreatic carboxypeptidase A.[43] This same group of investigators subsequently reported on the use of *N*-acetylimidazole for the determination of ''free'' tyrosyl residues in proteins[44] as opposed to ''buried'' residues. This has not necessarily proved to be the case.[45] The reaction of *N*-acetylimidazole with proteins has been well characterized.[46,47] *N*-acetylimidazole is commercially available but also can be easily synthesized.[39] Our laboratory generally synthesizes the reagent and always subjects reagent obtained from a commercial source to recrystallization from benzene after drying with sodium sulfate. It should be noted that, as with many reagents, *N*-acetylimidazole is hygroscopic and should be stored in a container, preferably a vacuum desiccator, over a suitable desiccant. A partial listing of proteins which have been modified with *N*-acetylimidazole is presented in Table 1. A stock solution of reagent is prepared in *dry* benzene (this stock solution is relatively stable for 2 to 4 weeks at 4°C) and a portion containing the desired amount of reagent introduced to the reaction vessel. The solvent (benzene) is removed by a stream of *dry* air or *dry* nitrogen. The reaction is initiated by the addition of the protein solution to be modified to the residue of reagent. The reaction is usually performed at pH 7.0 to 7.5. A wide variety of buffers has been used for the study of the reaction of *N*-acetylimidazole. A high concentration of nucleophilic species such as Tris should be avoided because of reagent instability.[44] Likewise, although the modification occurs more rapidly at pH values more alkaline than 7.5, reagent and product (*O*-acetyl tyrosine) stability becomes a significant problem.

There are several approaches to the determination of the extent of tyrosine modification by *N*-acetylimidazole. The amount of acetylhydroxamate produced by the reaction of hydroxylamine can be determined.[48] The procedure described by these investigators involves the addition of 0.25 mℓ of a hydroxylamine solution (4 M NH$_2$OH·HCl/3.5 M NaOH/0.001 M EDTA; 1/2/1) to 1.0 mℓ of the acetylated protein sample. After 1 min, 0.5 mℓ 25% trichloroacetic acid and 0.5 mℓ 20% FeCl$_2$, 6 H$_2$O in 2.5 M HCl is added and the absorbance of the supernatant fraction determined at 540 nm. We have found it convenient to use *p*-nitrophenyl acetate as the standard for this reaction. Secondly, *O*-acetylation of tyrosine produced a decrease in absorption at 278 nm. A $\Delta\epsilon$ = 1160 M^{-1} cm^{-1} has been reported[46] while a subsequent study reported a $\Delta\epsilon$ = 1210 M^{-1}cm^{-1}.[45] We have had more reliable results with the latter value in this laboratory. We have also found it more accurate to determine changes in absorbance at 278 nm as a function of time taking into account spectral changes introduced by the addition of reagent to a solvent blank.[45] One of the major advantages of reaction with *N*-acetylimidazole is the ease of reversal of the reaction. The *O*-acetyl derivative of tyrosine is unstable under mildly alkaline conditions, and presence of a nucleophile such as Tris greatly decreases the stability of the *O*-acetyl derivative. Quantitative deacetylation occurs with hydroxylamine at pH 7.5. As would be expected the rate of regeneration of free tyrosine is a function of hydroxylamine concentration. It should be noted that the primary side reaction products of the reaction of *N*-acetylimidazole with proteins, ϵ-*N*-acetyllysines and *N*-acetyl amino terminal amino acids, are stable to neutral or alkaline hydroxylamine. Assignment of changes in the biological activity of a protein on reaction with *N*-acetylimidazole to the *O*-acetylation of tyrosine can be verified by the reversibility of such changes in the presence of hydroxylamine.

The possible use of tetranitromethane for the modification of tyrosyl residues in proteins was advanced over 30 years ago.[49] This reagent was used shortly after for the modification of several proteins including thrombin.[50] However, it was not until some 2 decades later that the studies of Vallee, Riordan, and Sokolovsky established the specificity and characteristics of the reaction of tetranitromethane with proteins.[51,52]

The modification (Figure 6) proceeds optimally at alkaline pH (see Figure 7). The rate

Table 1
REACTION OF PROTEINS WITH *N*-ACETYLIMIDAZOLE

Protein	Solvent/temp	Reagent excess[a]	*O*–AcTyr/ Tyr[b]	Ref.
Carboxypeptidase[c]	0.02 *M* sodium barbital, 2.0 *M* NaCl, pH 7.5/23°C	60	4.3/19[d]	1
Pepsinogen	0.02 *M* sodium Veronal, 2.0 *M* NaCl, pH 7.5/25°C	60	7/16[e]	2
Pepsin	2.0 *M* NaCl, pH 5.8/25°C[f]	60	9/15[g]	2
Trypsin	0.01 *M* sodium borate, 0.01 *M* CaCl$_2$, pH 7.6/0°C	30	1.7/10[h]	3
Trypsin	0.01 *M* sodium borate,[i] 0.01 *M* CaCl$_2$, pH 7.6/0°C	465	3.0/10[j]	3
α-Amylase[k]	0.02 *M* Tris·Cl, pH 7.5/25°C	500	3.5/12[l]	4
Subtilisin novo	0.016 *M* barbital, pH 7.5/	100	7/10[m]	5
Subtilisin carlsberg	0.016 *M* barbital, pH 7.5/	130	8.4/13[m]	5
Hemerythrin	0.05 *M* sodium borate, 0.05 *M* Tris, pH 7.5/0°C	800	—[n]	6
Thrombin	0.02 *M* Tris, 0.02 *M* imidazole, 0.02 *M* acetate, pH 7.5/ 23°C	300	4.4/12	7
Fructose diphosphatase	0.050 *M* sodium borate, pH 7.5	—	—	8
Erythrocyte ATPase				
Stroma	0.010 *M* Tris, pH 7.4/23°C[o]	—	—	9
Intact cells	0.010 *M* Tris, 0.140 *M* NaCl, pH 7.4/23°C[p]	—	—	9
α-Lactalbumin		200	2/5[q]	10,11
Pancreatic colipase	—	—	—	12,13
Pancreatic α-amylase	0.01 *M* phosphate, pH 7.5[r], 0.1 m*M* CaCl/25°C	120	5.9/18	14
Sweet potato α-amylase	0.01 *M* acetate, pH 7.5/25°C[r]	120	5.3/17	14
Aspergillus niger glucamylase	0.01 *M* acetate, pH 7.5/25°C	120	11.3/33	14
Emulsin β-D-glucosidase	0.01 *M* phosphate, pH 6.1/ 25°C[r]	300	—	15

[a] Moles *N*-acetylimidazole/mole protein.

[b] Moles *O*-acetyltyrosine/moles tyrosine in modified protein.

[c] Bovine pancreatic carboxypeptidase A-Anson.

[d] Changes in catalytic activity reversed by treatment with 0.01 *M* hydroxylamine, pH 7.5 at 23°C. Primary amino groups were not acetylated under those reaction conditions.

[e] Five out of ten lysine residues modified.

[f] pH Maintained by NaOH from pH stat.

[g] Lysine not acetylated under these conditions. Reaction with 1.0 *M* hydroxylamine at pH 5.8 (60 min; 37°C) reversed changes in catalytic activity produced on reaction with *N*-acetylimidazole and presumably deacetylated *O*-acetyl tyrosyl residues.

[h] Also 1.0 serine and 0.3 lysine.

[i] Also used Tris, TES, HEPES, and barbital buffers without any significant difference in nature of the reaction.

[j] Also 1.7 (probably serine and histidine) and 2.5 lysine residues modified.

[k] From *Bacillus subtilis*.

[l] Approximately 2 lysine residues modified under these conditions. Only a single tyrosine residue is modified with tetranitromethane. Either reagent (tetranitromethane or *N*-acetylimidazole) led to a 70 to 80% loss of catalytic activity.

[m] The reaction with *N*-acetylimidazole was performed with subtilisin preparation previously treated with phenylmethanesulfonyl fluoride. The active enzyme catalyzes the rapid hydrolysis of *N*-acetylimidazole under reaction conditions.

[n] Reaction performed on protein where lysine residue had been previously blocked by reaction with ethyl acetimidate. *N*-acetylimidazole was added in 4 200-fold molar excess portion at 2-hr intervals.

Table 1 (continued)
REACTION OF PROTEINS WITH *N*-ACETYLIMIDAZOLE

[o] Reaction for 1 hr at ambient temperature with amount of *N*-acetylimidazole equivalent (weight/weight basis) to stroma. The reaction mixture was washed with distilled water to remove *N*-acetylimidazole.

[p] Reaction for 1 hr at ambient temperature. The quantity of *N*-acetylimidazole used is not given. It is stated that this reagent should readily pass across the cell membrane but this conclusion is based on analogy with acetic anhydride.

[q] Extensive modification of amino groups was reported.

[r] *N*-acetylimidazole added as a solid; pH maintained at 7.5 with pH stat.

References for Table 1

1. **Simpson, R. T., Riordan, J. F., and Vallee, B. L.,** Functional tyrosyl residues in the active center of bovine pancreatic carboxypeptidase A, *Biochemistry,* 2, 616, 1963.
2. **Perlmann, G. E.,** Acetylation of pepsin and pepsinogen, *J. Biol. Chem.,* 241, 153, 1966.
3. **Houston, L. L. and Walsh, K. A.,** The transient inactivation of trypsin by mild acetylation with *N*-acetylimidazole, *Biochemistry,* 9, 156, 1970.
4. **Connellan, J. M. and Shaw, D. C.,** The inactivation of *Bacillus subtilis* α-amylase by *N*-acetylimidazole and tetranitromethane. Reaction of tyrosyl residues, *J. Biol. Chem.,* 245, 2845, 1970.
5. **Myers, B., II and Glazer, A. N.,** Spectroscopic studies of the exposure of tyrosine residues in proteins with special reference to the subtilisins, *J. Biol. Chem.,* 246, 412, 1971.
6. **Fan, C. C. and York, J. L.,** The role of tyrosine in the hemerythrin active site, *Biochem. Biophys. Res. Commun.,* 47, 472, 1972.
7. **Lundblad, R. L., Harrison, J. H., and Mann, K. G.,** On the reaction of purified bovine thrombin with *N*-acetylimidazole, *Biochemistry,* 12, 409, 1973.
8. **Kirtley, M. E. and Dix, J. C.,** Effects of acetylimidazole on the hydrolysis of fructose diphosphate and *p*-nitrophenyl phosphate by liver fructose diphosphatase, *Biochemistry,* 13, 4469, 1974.
9. **Masiak, S. J. and D'Angelo, G.,** Effects of *N*-acetylimidazole on human erythrocyte ATPase activity. Evidence for a tyrosyl residue at the ATP binding site of the (Na^+, K^+)-dependent ATPase, *Biochim. Biophys. Acta,* 382, 83, 1975.
10. **Kronman, M. J., Hoffman, W. B., Jeroszko, J., and Sage, G. W.,** Inter and intramolecular interactions of α-lactalbumin. XI. Comparison of the "exposure" of tyrosyl, tryptophyl and lysyl side chains in the goat and bovine proteins, *Biochim. Biophys. Acta,* 285, 124, 1972.
11. **Holohan, P., Hoffman, W. B., and Kronman, M. J.,** Chemical modification of tyrosyl and lysyl residues in goat alpha lactalbumin and the effect on the interaction with the galactosyl transferase, *Biochim. Biophys. Acta,* 621, 333, 1980.
12. **Erlanson, C., Barrowman, J. A., and Borgström, B.,** Chemical modifications of pancreatic colipase, *Biochem. Biophys. Acta,* 489, 150, 1977.
13. **Erlanson-Albertsson, C.,** The importance of the tyrosine residues in pancreatic colipase for its activity, *FEBS Lett.,* 117, 295, 1980.
14. **Hoschke, A., Laszlo, E., and Hollo, J.,** A study of the role of tyrosine groups at the active centre of amylolytic enzymes, *Carbohydrate Res.,* 81, 157, 1980.
15. **Kiss, L., Korodi, I., and Nanasi, P.,** Study on the role of tyrosine side-chains at the active center of emulsin β-D-glucosidase, *Biochim. Biophys. Acta,* 662, 308, 1981.

of modification of *N*-acetyltyrosine is twice as rapid at pH 8.0 as at pH 7.0; it is approximately ten times as rapid at pH 9.5 as at pH 7.0. As shown above, the reaction of tetranitromethane with tyrosine produces 3-nitrotyrosine, nitroformate, and two protons. The spectral properties of nitroformate (ε at 350 nm = 14,400) suggested that monitoring the formation of this species would be a sensitive method for monitoring the time course of the reaction of tetranitromethane with tyrosyl residues.[51] Although determining the rate of nitroformate production appears to be effective in studying the reaction of tetranitromethane with model compounds such as *N*-acetyltyrosine (Figure 8), it has not proved useful with proteins.[53,54] Although the reaction of tetranitromethane with proteins is reasonably specific for tyrosine, oxidation of sulfhydryl groups (Figure 9) has been reported[53,54] as has reaction with histidine,[53] methionine,[53] and tryptophan.[53,55] Reaction with histidine and tryptophan is, however,

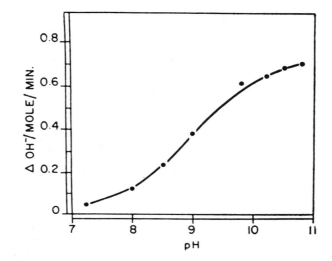

FIGURE 6.   A scheme for the modification of tyrosyl residues with tetranitromethane.

FIGURE 7.   Dependence of the rate of nitration of *N*-acetylty-rosine on pH. Rates were calculated from the initial linear slope of the titration curves when nitrations were performed on the pH-stat. (From Sokolovsky, M., Riordan, J. F., and Vallee, B. L., *Biochemistry*, 5, 3582, 1966. With permission.)

unusual and, in general, care need be taken only with respect to potential reaction with sulfhydryl groups. The reaction of tetranitromethane with 2-keto-4-hydroxyglutarate aldolase[56] provides a particularly good example of sulfhydryl modification. Figure 10 describes the inactivation of the enzyme as a function of tetranitromethane concentration in 0.05 *M* Tris, pH 8.0 at 20°C. Cysteinyl residues are lost as a result of this modification (Figure 11). Under conditions where 4 cysteinyl residues are modified in the enzyme, only 0.4 mol of 3-nitrotyrosine is formed per mole of enzyme (Figure 12). The relationship between cysteinyl residue modification and tetranitromethane concentration is shown in Figure 13. The loss of cysteine reflected *both* cystine formation and oxidation to cysteic acid. It is noted that significant reaction with model compounds containing histidine, methionine, or tryptophan was observed only at pH values greater than 8.[53]

   The other potential problem associated with the use of tetranitromethane for the modification of tyrosyl residues in proteins is the covalent cross-linkage of tyrosyl residues resulting in inter- and intramolecular association. The magnitude of this problem is dependent on variables such as protein concentration and solvent conditions (i.e., pH). With respect to this latter consideration it is noted that acidification of reaction mixtures tends to favor the cross-linkage reaction.[54] As would be expected, the extent of cross-linkage observed varies with the protein being studied. For example, reaction of pancreatic deoxyribonuclease with tetranitromethane results in extensive formation of dimer.[57] Polymerization of protein on

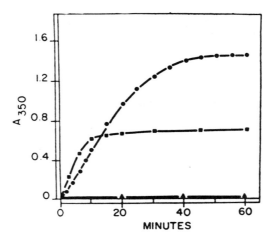

FIGURE 8.   Time course of formation of nitroformate on reaction of tetranitromethane with several amino acid derivatives. Shown is the increase in absorbance at 350 nm on the reaction of tetranitromethane with *N*-acetyltyrosine (●), glutathione (■), and *N*-carbobenzoxyglycyl-L-tryptophan, *N*-acetylhistidine, or *N*-carbobenzoxy-L-methionylglycine (▲), all 0.1 m*M*. Tetranitromethane (5 μℓ, 42 μmol) was added to 3 mℓ 0.05 *M* Tris, pH 8.0, containing the amino acid derivative at 20°C. The data are corrected for the absorbance due to *N*-acetyl-3-nitrotyrosine. (From Sokolovsky, M., Riordan, J. F., and Vallee, B. L., *Biochemistry*, 5, 3582, 1966. With permission.)

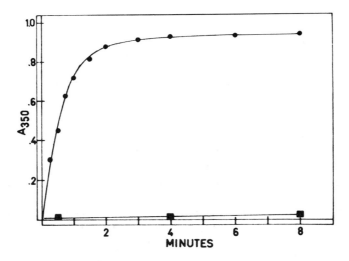

FIGURE 9.   The formation of nitroformate on the reaction of tetranitromethane with reduced or oxidized glutathione. Shown is the increase in A_{350} on the addition of a tenfold molar excess of tetranitromethane to either 100 μ*M* reduced glutathione (●) or 100 μ*M* oxidized glutathione (■) in 0.1 *M* acetate, pH 5.5, at 20°C. (From Sokolovsky, M., Harrell, D., and Riordan, J. F., *Biochemistry*, 8, 4740, 1969. With permission.)

treatment with tetranitromethane can be effectively evaluated by gel filtration as shown in Figures 14 to 16. The experiments described in Figure 16[60] deserve further comment. The modification of phospholipase A_2 by tetranitromethane is accomplished in 0.050 *M* Tris, 0.1 *M* NaCl, 0.010 *M* $CaCl_2$, pH 8.0 with a tenfold molar excess of tetranitromethane. The

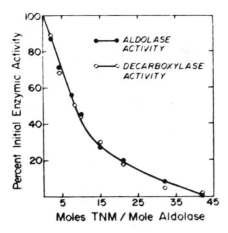

FIGURE 10.   The reaction of 2-keto-4-hydroxyglutarate aldolase (KHG-aldolase) with tetranitromethane. Shown is the inactivation of aldolase and β-decarboxylase activities by
varying molar excesses of tetranitromethane. KHG-aldolase (0.24 mg) was incubated in 0.20
mℓ of 0.05 M Tris-HCl buffer (pH 8.0) for 30 min at 20°C with the indicated concentrations
of tetranitromethane. Portions were withdrawn, diluted 50-fold with the same buffer at 0°C,
and assayed for aldolase and β-decarboxylase activities. (From Lane, R. S. and Dekker, E.
E., *Biochemistry,* 11, 3295, 1972. With permission.)

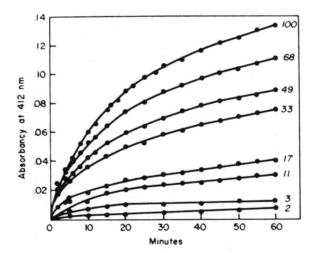

FIGURE 11.   Kinetics of the reaction of 5,5′-dithiobis(2-nitrobenzoic acid) with native and
tetranitromethane-modified 2-keto-4-hydroxyglutarate aldolase (KHG-aldolase). Titrations
were carried out at 25°C in 1.0 mℓ of 0.05 M Tris-HCl buffer (pH 8.0) with 0.5 mM 5,5′-
dithiobis(2-nitrobenzoic acid). The protein concentration was approximately 0.4 mg/mℓ. The
values listed in the right-hand margin rpresent the percent initial KHG-aldolase activity
remaining after the enzyme (1.04 mg) was incubated in 0.8 mℓ of 0.05 M Tris-HCl buffer
(pH 8.0) for 30 min at 20°C with increasing amounts of tetranitromethane (molar excess 2-
to 39-fold). The protein solutions were exhaustively dialyzed against 0.05 M Tris-HCl buffer
(pH 8.0) at 4°C prior to titration with 5,5′-dithiobis (2-nitrobenzoic acid). (From Lane, R.
S. and Dekker, E. E., *Biochemistry,* 11, 3295, 1972. With permission.)

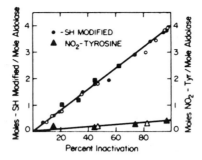

FIGURE 12.   Relationship of tyrosine and cysteine modification to the loss of catalytic activity of 2-keto-4-hydroxyglutarate aldolase (KHG-aldolase) on reaction with tetranitromethane. Shown is the extent of tetranitromethane inactivation as a function of the number of modified cysteinyl and tyrosyl residues. Experimental conditions for the attainment of different levels of inactivation were the same as outlined in the caption to Figure 11. After reaction for 30 min at 20°C, the protein solutions were exhaustively dialyzed against 0.05 *M* Tris-HCl buffer (pH 8.0) at 4°C. The dialyzed solutions were used to determine protein content, aldolase activity (closed symbols), β-decarboxylase activity (open symbols), sulfhydryl content (○,●), and nitrotyrosyl content (△,▲). The difference in sulfhydryl content between the native and tetranitromethane-inactivated KHG-aldolase was taken as the number of sulfhydryl groups modified; (■), the number of thiol groups modified when the reaction was carried out in the presence of 10 m*M* DL-2-keto-4-hydroxyglutarate(KHG). (From Lane, R. S. and Dekker, E. E., *Biochemistry,* 11, 3295, 1972. With permission.)

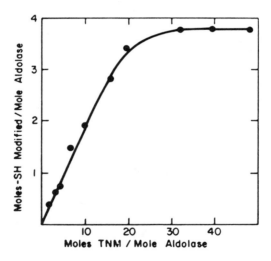

FIGURE 13.   Extent of sulfhydryl group modification in 2-keto-4-hydroxyglutarate aldolase as a function of tetranitromethane (TNM) concentration. Reaction conditions were as described in the caption to Figure 11. After incubation for 30 min at 20°C, the protein solutions were dialyzed exhaustively against 0.05 *M* Tris-HCl buffer (pH 8.0) at 4°C and sulfhydryl groups were determined by titration with 5,5'-dithiobis(2-nitrobenzoic acid). (From Lane, R. S. and Dekker, E. E., *Biochemistry,* 11,3295, 1972. With permission.)

time course for the inactivation of enzyme obtained from either (A) equine, (B) porcine, (C) bovine sources is shown in Figure 17. When the reaction is performed in the absence of substrate (egg yolk lysolecithin micelles shown in solid symbols in Figure 17), significant polymerization is observed (Figure 16, panel A). When substrate is included, the rate of

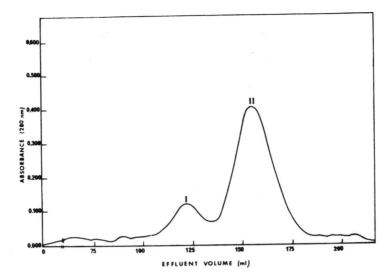

FIGURE 14. Typical separation by gel filtration of *Mucor miehei* protease after reaction with tetranitromethane. Modification of the enzyme was performed at 15°C in 0.1 *M* phosphate buffer (pH 8.0). Gel filtration was performed on G-75 Sephadex (2.5 × 90 cm column). Peak I is the minor polymeric reaction product; peak II is the major reaction product with the same chromatographic mobility as the native enzyme. (From Rickert, W. S. and McBride-Warren, P. A., *Biochim. Biophys. Acta*, 371, 368, 1974. With permission.)

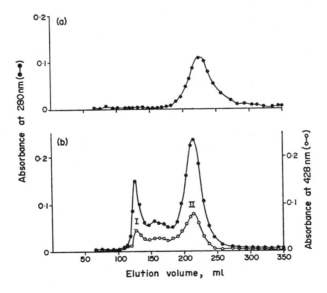

FIGURE 15. The gel filtration of α_1-antitrypsin before and after reaction with tetranitromethane. Shown are the elution profiles of gel filtration on Sephadex G-100 column (2.5 × 80 cm) in 0.05 *M* Tris-HCl buffer, pH 7.5 containing 0.1 *M* KCl and 0.02% sodium azide. The elution was carried out at ambient temperature. (a) Native α_1-antitrypsin. (b) After nitration of α_1-antitrypsin: 15 mg of the nitrated α_1-antitrypsin were applied to the column. The recovery was 2.4 mg in peak I(polymeric product) and 9.4 mg in peak II(monomeric form). The α_1-antitrypsin(1 mg/mℓ) was nitrated at a molar ratio of tyrosine to tetranitromethane of 1:17.2 in 0.05 *M* Tris-HCl, pH 8.0, at 25°C. (From Busby, T. F. and Gan, J. C., *Int. J. Biochem.*, 6, 835, 1975. With permission.)

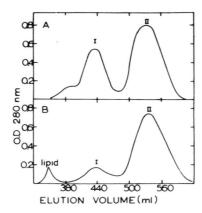

ELUTION VOLUME (ml)

FIGURE 16.   Gel filtration patterns of nitrated horse phospholipase A_2 in 1% ammonium
bicarbonate, pH 8.0, from a column (3.5 × 300 cm) of Sephadex G-75. The following are
the experimental conditions: equine phospholipase A_2 was nitrated in the (A) absence and
(B) presence of egg yolk lysolecithin used as the sample. Peaks I represent dimeric enzymes;
peaks II are monomeric enzymes. (From Meyer, H., Verhoef, H., Hendriks, F. F. A.,
Slotboom, A. J., and de Haas, G. H., *Biochemistry*, 8, 3582, 1979. With permission.)

inactivation is increased (open symbols in Figure 17) and the extent of polymerization
decreases (Figure 16, panel B). It is suggested that the increased rate of inactivation reflects
the increased solubility of tetranitromethane in the apolar interior of the substrate micelle
enhancing reaction with enzyme bound to the micelle. The binding of phospholipase to the
micelle reduces polymerization, which probably only occurs with enzyme in solution.

   The extent of modification of tyrosyl residues by tetranitromethane in proteins can be
assessed by either spectophotometric means or by amino acid analysis. At alkaline pH (pH
≥8), 3-nitrotyrosine has an absorption maximum at 428 nm (Figure 18) with ϵ = 4100
M^{-1} cm^{-1}; the absorption maximum of tyrosine at 275 nm increases from ϵ = 1360 M^{-1}
cm^{-1} to 4000 M^{-1} cm^{-1}. At acid pH (pH ≤ 6), the absorption maximum is shifted from
428 nm to 360 nm with an isosbestic point at 381 nm (ϵ = 2200 M^{-1} cm^{-1}) (see Figure
19). We have found it convenient to determine the A_{428} in 0.1 M NaOH. Amino acid analysis
after acid hydrolysis has also proved to be a convenient method of assessing the extent of
3-nitrotyrosine formation. 3-Nitrotyrosine is stable to acid hydrolysis (6 N HCl/105°C/24
hr).[52] This approach has the added advantage that other modifications of tyrosine such as
free radical-mediated cross-linkage can be either excluded or quantitatively determined. If
nitration to form 3-nitrotyrosine is the only modification of tyrosyl residues in a protein
occurring on reaction with tetranitromethane, the sum of 3-nitrotyrosine and tyrosine should
be equivalent to the amount of tyrosine in the unmodified protein.

   There are several consequences of the nitration of a tyrosyl residue. The most obvious is
the placing of a somewhat bulky substituent (the nitro group) *ortho* to the phenolic hydroxyl
function. The properties of the substituent nitro group ''push'' electrons into the benzene
ring (inductive effect), lowering the pKa of the phenolic hydroxyl from approximately 10.3
to approximately 7.3. This of course means that the phenolic hydroxyl of the nitrated tyrosyl
residue will be in a partially ionized state at physiological pH. The nitro function can be
reduced to the corresponding amine under relatively mild conditions ($Na_2S_2O_4$, 0.05 M Tris,
pH 8.0).[61] The conversion of 3-nitrotyrosine to 3-aminotyrosine is associated with loss of
the absorption maximum at 428 nm (Figure 20) and the change in the pKa of the phenolic
hydroxyl group from approximately 7.0 to 10.0. The resultant amine function can be subsequently modified.[62]

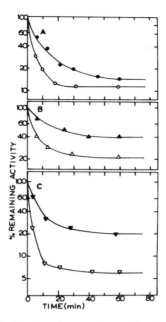

FIGURE 17. The modification of equine, porcine, and bovine phospholipase A₂ by tetra-nitromethane in the presence or absence of egg yolk lysolecithin. Shown is the loss of phospholipase A₂ activity as a function of time upon addition of tetranitromethane. The following are the experimental conditions. To a solution(2 mℓ) of (A) equine, (B) porcine, and (C) bovine phospholipase A₂(1 mg/mℓ) in a buffer containing 0.1 M NaCl, 10 mM CaCl₂, and 50 mM Tris-HCl, pH 8.0, was added 20 μℓ of a solution of 1% tetranitromethane in ethanol. Incubation was performed at 30°C. At suitable time intervals, portions (25 to 200 μℓ) were withdrawn for the determination of enzymatic activity; (●) equine, (▲) porcine, and (▼) bovine phospholipases. Similar incubations were made in the presence of egg yolk lysolecithin(10 mg/mℓ): (○) equine, (△) porcine, and (▽) bovine phospholipases. For the bovine enzyme in the presence of lysolecithin, 50 mM CaCl₂ was used. (From Meyer, H., Verhoef, H., Hendriks, F. F. A., Slotboom, A. J., and de Haas, G. H., *Biochemistry,* 18, 3582, 1979. With permission.)

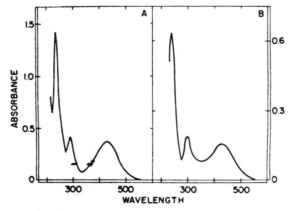

FIGURE 18. UV absorption spectra of *N*-acetyl-3-nitrotyrosine and nitrocarboxypeptidase. Shown is (A), the absorption spectrum of *N*-acetyl-3-nitrotyrosine minus *N*-acetyltyrosine, both at 0.100 mM. (B) The absorption spectrum of nitrocarboxypeptidase minus carboxypeptidase both at 17.6 μM.. Both spectra were obtained in 0.05 M Tris-1 M NaCl, pH 8.0, at ambient temperature. (From Riordan, J. F., Sokolovsky, M., and Vallee, B. L., *Biochemistry,* 6, 3609, 1967. With permission.)

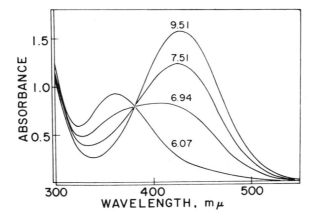

FIGURE 19.   pH Dependence of the UV absorption spectra of the 3-nitro derivative of tyrosine. Shown are the absorption spectra of *N*-acetyl-3-nitrotyrosine (250 μ*M*) in 0.2 *M* Tris, 0.2 *M* acetate, 0.5 *M* NaCl at the pH indicated. (From Riordan, J. F., Sokolovsky, M., and Vallee, B. L., *Biochemistry*, 6, 358, 1967. With permission.)

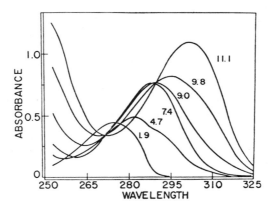

FIGURE 20.   pH Dependence of the UV absorption spectra of the 3-amino derivative of tyrosine. Shown are the absorption spectra of 267 μ*M* 3-aminotyrosine in 0.2 *M* Tris, 0.2 *M* acetate, 0.5 *M* NaCl at the pH value indicated. 3-Aminotyrosine was prepared by the reaction of 3-nitrotyrosine with a 6-fold molar excess of sodium hydrosulfite ($Na_2S_2O_2$) in 0.05 *M* Tris, pH 8.0. (From Sokolovsky, M., Riordan, J. F., and Vallee, B. L., *Biochem. Biophys. Res. Commun.*, 27, 20, 1967. With permission.)

In addition to changing the properties of a given tyrosyl residue, nitration also introduces a spectral probe which can be used to detect conformational change in the protein. The concept of using reagents to introduce probes with unique spectral and fluorescent properties has been introduced in Volume I, Chapter 1. 3-Nitrotyrosine, as mentioned above, has an absorption maximum at 428 nm at alkaline pH. This spectral property was first used by Riordan and co-workers with studies on nitrated carboxypeptidase A[63] to study changes in the microenvironment around the modified residue. Addition of β-phenylpropionate, a com-

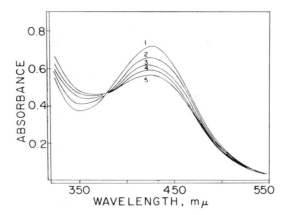

FIGURE 21. The effect of β-phenylpropionate, a competitive inhibitor of carboxypeptidase, on the UV absorption
spectrum of nitrocarboxypeptidase. Spectra were measured
in 0.2 M Tris, 0.2 M acetate, 0.5 M NaCl, pH 8.0. The
concentrations of β-phenylpropionate were as follows:
1(control), none; 2, 0.01 M; 3, 0.025 M; 4, 0.05 M; and
5, 0.1 M. (From Riordan, J. F., Sokolovsky, M., and Vallee, B. L., *Biochemistry*, 6, 358, 1967. With permission.)

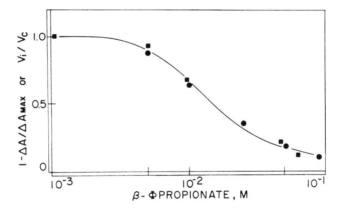

FIGURE 22. The effect of a competitive inhibitor on the catalytic and
spectral properties of nitrocarboxypeptidase. Shown is the effect of β-
phenylpropionate on the esterase activity (■) and on the absorbance at
428 nm (●) of nitrocarboxypeptidase. $\Delta A/\Delta A_{max}$ represents the fractional
decrease in absorbance at 428 nm observed in the presence of the concentration of β-phenylpropionate indicated, while ΔA_{max} is the maximal
decrease in absorbance at 428 nm calculated by extrapolation to infinite
concentration of β-phenylpropionate. (From Riordan, J. F., Sokolovsky,
M., and Vallee, B. L., *Biochemistry*, 6, 358, 1967. With permission.)

petitive inhibitor of carboxypeptidase and nitrated carboxypeptidase, decreased the absorbance of mononitrocarboxypeptidase at 428 nm (Figure 21). This change is consistent with
an increase in the hydrophobic quality of the microenvironment surrounding the modified
tyrosyl residue. There was a direct correlation between inhibition of esterase activity by β-
phenylpropionate and the decrease in absorbance at 428 nm (Figure 22).

Other uses of tetranitromethane in the study of the relationship between structure and
function in various proteins are presented in Table 2. The methods used in the various studies
are closely related to those described above.

Table 2
THE USE OF TETRANITROMETHANE TO MODIFY TYROSYL RESIDUES IN PROTEINS

Protein concentration	Solvent/temp	Molar excess TNM	Residues modified	Ref.
Carboxypeptidase A (10 mg/mℓ)	0.05 M Tris, 2 M NaCl/20°C	4	1.2/18	1
Staphylococcal nuclease (2 mg/mℓ)	0.05 M Tris, pH 8.1 (23°C)	2	1.1/7	2
		4	1.6/7	
		8	3.4/7	
		12	3.7/7	
		16	4.2/7	
		20	4.7/7	
		30	5.0/7	
		60	4.8/7	
	4 M guanidine	60	6.4/7	
Horse heart cytochrome C (1 mM)	0.05 M Tris, pH 8.0	16	2/4	3
Aspartate aminotransferase (5 mg/mℓ)	0.05 M Tris, pH 7.5 (22°C)	30	—	4
Thrombin (0.06 mg/mℓ)	0.03 M sodium phosphate, pH 8.0 (24°C)	1000	4.9/12	5
2-keto-4-hydroxyglutarate aldolase (1.2 mg/mℓ)	0.05 M Tris, pH 8.0			6
Porcine carboxypeptidase-β (5—7 mg/mℓ)	0.05 M Tris, pH 8.0 (23°C)	8	1.2/21	7
Bovine pituitary growth hormone (4 mg/mℓ)	0.05 M Tris, pH 8.0 (0°C)	30	2.7/6	8
Ovine pituitary growth hormone (4 mg/mℓ)	0.05 M Tris, pH 8.0 (0°C)	30	3.0/6	8
Aspartate transcarbamylase (catalytic subunit, 4 mg/mℓ)	0.1 M potassium phosphate, pH 6.7 (23°C)	2/3/8		9
Mucor miehei protease (1 mg/mℓ)	0.1 M phosphate, pH 8.0 (15°C)	20	2.8/21	10
Turnip yellow mosaic virus capsids (1.75 mg/mℓ)	0.05 M Tris, pH 8.0 (22°C)	50	3/3	11
α_1-Antiprotease inhibitor (1 mg/mℓ)	0.05 M Tris, pH 8.0 (25°C)	120	3/7	12
	5 M guanidine (25°C)	60	7/7	12
α_1-Acid glycoprotein	0.1 M Tris, pH 8.0 (23°C)	10	2.7/12	13
Carboxypeptidase A crystals (5 mg/mℓ)	0.05 M Tris, pH 8.0 (20°C)		1/9	14
Bovine growth hormone (2 mg/mℓ)	0.03 M Ringer phosphate (25°C)	12	3/7	15
Equine growth hormone (2 mg/mℓ)	0.03 M Ringer phosphate (25°C)	12	3/7	15
Lactose repressor protein	0.1 M Tris, 0.1 M mannose, pH 7.8 (23°C)	800	2.4/8	16
Aspartate transcarbamylase (8.3 mg/mℓ)	0.1 M Tris, acetate (25°C)	750	2.2/10	17
Aspartate transcarbamylase (5 mg/mℓ)	0.1 M Tris, pH 8.0 (25°C)		2.2/10	18
Human serum albumin	pH 8.0	80	9/18	19,20
			1.2/18	
Prolactin (1 mg/mℓ)	0.05 M Tris, pH 8.0 (23°C)	175	1.9/7	21
Porcine pancreatic phospholipase (1 mg/mℓ)	0.05 M Tris, 0.1 M NaCl, 0.01 M CaCl$_2$, pH 8.0 (30°C)	10	—	22
Equine pancreatic phospholipase (1 mg/mℓ)	0.05 M Tris, 0.1 M NaCl, 0.01 M CaCl$_2$, pH 8.0 (30°C)	10	—	22

Table 2 (continued)
THE USE OF TETRANITROMETHANE TO MODIFY TYROSYL RESIDUES IN PROTEINS

Protein concentration	Solvent/temp	Molar excess TNM	Residues modified	Ref.
Bovine pancreatic phospholi-pase (1 mg/mℓ)	0.05 M Tris, 0.1 M NaCl, 0.01 M CaCl$_2$, pH 8.0 (30°C)	10	—	22
Troponin C (1 mg/mℓ)	0.05 M Tris, 0.002 M EGTA (23°C)	8	3/3	23
Mouse myeloma protein (5 × 10^{-5} M)	0.01 M Tris, pH 8.2 (23°C)	10		24
Escherichia coli elongation factor G (4—6 mg/mℓ)	0.1 M Tris, 0.01 M KCl, 5% glycerol, 0.2 mM EDTA, pH 8.0 (25°C)	250	4/20	25
Elapid venom cardiotoxins (7 mg/mℓ)	0.1 M Tris, pH 7.0 (25°C) or 0.05 M Tris, pH 8.0 (25°C)	—	—	26
Lactose repressor (0.1—1.0 mg/mℓ)	0.1 M Tris, pH 8.0 or 0.24 M potassium phosphate, 5% glucose, pH 8.0 (23°C)	50	—	27
L-Lactate monooxygenase (1.8 μM)	0.05 M Tris, pH 8.0, 7.5 (30°C)			28
Tryptophanase apoenzyme (0.1 μM)	0.05 M triethanolamine pH 8.0 (30°C)	—	—	29
β-Lactamase (1.3 mg/mℓ)	0.05 M Tris, pH 8.0 (25°C)	5.20		30

References for Table 2

1. **Riordan, J. F., Sokolovsky, M., and Vallee, B. L.,** The functional tyrosyl residues of carboxypeptidase A. Nitration with tetranitromethane, *Biochemistry,* 6, 3609, 1967.
2. **Cuatrecasas, P., Fuchs, S., and Anfinsen, C. B.,** The tyrosyl residues at the active site of staphylococcal nuclease. Modifications by tetranitromethane, *J. Biol. Chem.,* 243, 4787, 1968.
3. **Skov, K., Hofmann, T., and Williams, G. R.,** The nitration of cytochrome c, *Can. J. Biochem.,* 47, 750, 1969.
4. **Christen, P. and Riordan, J. F.,** Syncatalytic modification of a functional tyrosyl residue in aspartate aminotransferase, *Biochemistry,* 9, 3025, 1970.
5. **Lundblad, R. L. and Harrison, J. H.,** The differential effect of tetranitromethane on the proteinase and esterase activity of bovine thrombin, *Biochem. Biophys. Res. Commun.,* 45, 1344, 1971.
6. **Lane, R. S. and Dekker, E. E.,** Oxidation of sulfhydryl groups of bovine liver 2-keto-4-hydroxyglutarate aldolase by tetranitromethane, *Biochemistry,* 11, 3295, 1972.
7. **Sokolovsky, M.,** Porcine carboxypeptidase B. Nitration of the functional tyrosyl residue with tetranitromethane, *Eur. J. Biochem.,* 25, 267, 1972.
8. **Glaser, C. B., Bewley, T. A., and Li, C. H.,** Reaction of bovine and ovine pituitary growth hormones with tetranitromethane, *Biochemistry,* 12, 3379, 1973.
9. **Kirschner, M. W. and Schachman, H. K.,** Conformational studies on the nitrated catalytic subunit of aspartate transcarbamylase, *Biochemistry,* 12, 2987, 1973.
10. **Rickert, W. S. and McBride-Warren, P. A.,** Structural and functional determinants of *Mucor miehei* protease. IV. Nitration and spectrophotometric titration of tyrosine residues, *Biochim. Biophys. Acta,* 371, 368, 1974.
11. **Re, G. G. and Kaper, J. M.,** Chemical accessibility of tyrosyl and lysyl residues in turnip yellow mosaic virus capsids, *Biochemistry,* 14, 4492, 1975.
12. **Busby, T. F. and Gan, J. C.,** The reaction of tetanitromethane with human plasma α$_1$-antitrypsin, *Int. J. Biochem.,* 6, 835, 1975.
13. **Kute, T. and Westphal, U.,** Steroid-protein interactions. XXXIV. Chemical modification of α$_1$-acid glycoprotein for characterization of the progesterone binding site, *Biochim. Biophys. Acta,* 420, 195, 1976.
14. **Muszynska, G. and Riordan, J. F.,** Chemical modification of carboxypeptidase A crystals. Nitration of tyrosine 248, *Biochemistry,* 15, 46, 1976.
15. **Daurat-Larroque, S. T., Portuguez, M. E. M., and Santome, J. A.,** Reaction of bovine and equine growth hormones with tetranitromethane, *Int. J. Peptide Protein Res.,* 9, 119, 1977.

Table 2 (continued)

16. **Alexander, M. E., Burgum, A. A., Noall, R. A., Shaw, M. D., and Matthews, K. S.,** Modification of tyrosine residues of the lactose repressor protein, *Biochim. Biophys. Acta,* 493, 367, 1977.
17. **Landfear, S. M., Lipscomb, W. N., and Evans, D. R.,** Functional modifications of aspartate transcarbamylase induced by nitration with tetranitromethane, *J. Biol. Chem.,* 253, 3988, 1978.
18. **Lauritzen, A. M., Landfear, S. M., and Lipscomb, W. N.,** Inactivation of the catalytic subunit of aspartate transcarbamylase by nitration with tetranitromethane, *J. Biol. Chem.,* 255, 602, 1980.
19. **Malan, P. G. and Edelhoch, H.,** Nitration of human serum albumin and bovine and human goiter thyroglobulins with tetranitromethane, *Biochemistry,* 9, 3205, 1970.
20. **Moravek, L., Saber, M. A., and Meloun, B.,** Steric accessibility of tyrosine residues in human serum albumin, *Collect. Czech. Chem. Commun.,* 44, 1657, 1979.
21. **Andersen, T. T., Zamierowski, M. M., and Ebner, K. E.,** Effect of nitration on prolactin activities, *Arch. Biochem. Biophys.,* 192, 112, 1979.
22. **Meyer, H., Verhoef, H., Hendriks, F. F. A., Slotboom, A. J., and de Haas, G. H.,** Comparative studies of tyrosine modification in pancreatic phospholipases. I. Reaction of tetranitromethane with pig, horse, and ox phospholipases A$_2$ and their zymogens, *Biochemistry,* 18, 3582, 1979.
23. **McCubbin, W. D., Hincke, M. T., and Kay, C. M.,** The utility of the nitrotyrosine chromophore as a spectroscopic probe in troponin C and modulator protein, *Can. J. Biochem.,* 57, 15, 1979.
24. **Gavish, M., Neriah, Y. B., Zakut, R., Givol, D., Dwek, R. A., and Jackson, W. R. C.,** On the role of Tyr 34$_L$ in the antibody combining site of the dinitrophenyl binding protein 315, *Mol. Immunol.,* 16, 957, 1979.
25. **Alakhov, Y. B., Zalite, I. K., and Kashparov, I. A.,** Tyrosine residues in the C-terminal domain of the elongation factor G are essential for its interaction with the ribosome, *Eur. J. Biochem.,* 105, 531, 1980.
26. **Carlsson, F. H. H.,** The preparation of 3-nitrotyrosyl derivatives of three elapid venom cardiotoxins, *Biochim. Biophys. Acta,* 624, 460, 1980.
27. **Hsieh, W.-T. and Matthews, K. S.,** Tetranitromethane modification of the tyrosine residues of the lactose repressor, *J. Biol. Chem.,* 256, 4856, 1981.
28. **Durfor, C. N. and Cromartie, T. H.,** Inactivation of L-lactate monooxygenase by nitration with tetranitromethane, *Arch. Biochem. Biophys.,* 210, 710, 1981.
29. **Nihira, T., Toraya, T., and Fukui, S.,** Modification of tryptophanase with tetranitromethane, *Eur. J. Biochem.,* 119, 273, 1981.
30. **Wolozin, B. L., Myerowitz, R., and Pratt, R. F.,** Specific chemical modification of the readily nitrated tyrosine of the R$_{TEM}$ β-lactamase and of *Bacillus cereus* β-lactamase. I. The role of this tyrosine in β-lactamase catalysis, *Biochim. Biophys. Acta,* 701, 153, 1982.

As would be expected, each of these studies produced unique observations worthy of further consideration. It is, however, not possible to describe each in further detail. It should be noted that *this is not* a comprehensive listing of all studies during the past 15 years.

One of the first proteins to be subjected to study with tetranitromethane was pancreatic carboxypeptidase A.[63] The data were interpreted to suggest that nitration of a single tyrosine residue was responsible for an increase in observed esterase activity and a decrease in peptidase activity. It was subsequently demonstrated that polymerization was a significant side reaction under the conditions used. Later studies[64] utilized bovine pancreatic carboxypeptidase A crystals to obviate polymerization. It has been demonstrated that tyrosine–248 is modified under these conditions.[64] This is the same residue which is modified with diazotized arsanilic acid.[10-12] The study of the spectral properties of nitrated carboxypeptidase,[63] discussed above, provided initial impetus for the use of 3-nitrotyrosine as a "reporter group".

The study of the reaction of tetranitromethane with staphylococcal nuclease[55] is of particular interest for several reasons. First, this study demonstrated that tetranitromethane could modify a tryptophan residue in this protein following denaturation. There were distinctly different patterns of reaction (in terms of both extent of modification and residues modified) in the presence and absence of deoxythymidine 3′,5′-diphosphate, a competitive inhibitor, and the ribonuclease activity of the enzyme was affected to a greater degree than the deoxyribonuclease activity (Figure 23). Nitration of one tyrosine residue markedly affected the nitration of a second tyrosyl residue, and a substantial increase in the enzymatic activity of the modified enzyme was obtained by reduction with sodium dithionite under conditions

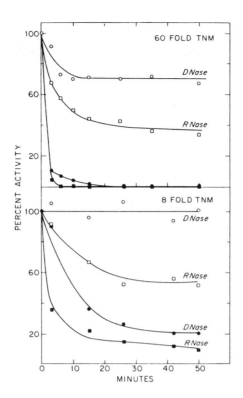

FIGURE 23.   The effect of reaction with tetranitromethane (TNM) on the catalytic activity(DNase and RNase) of Staphylococcal nuclease. Shown is the effect of nitration with either a 60-fold (top panel) or an 8-fold (bottom panel) molar excess of tetranitromethane on DNase and RNase activities of 0.12 mM Staphylococcal nuclease in the presence ($\circ$,$\square$) or absence ($\bullet$,$\blacksquare$) of 0.4 deoxythymidine 3′,5′-diphosphate and 10 mM CaCl$_2$. The reactions were performed in 0.05 M Tris-HCl, pH 8.1, at ambient temperature and portions were removed for enzyme assays at the indicated times. (From Cuatrecasas, P., Fuchs, S., and Anfinsen, C. B., *J. Biol. Chem.*, 243, 4787, 1968. With permission.)

similar to those described above. Selective substitution at the resulting amine group could be achieved, reflecting the low pKa (4.75). In this study, while reaction of dansyl chloride with the native protein at pH 8.2 resulted in random modification at a number of lysine residues, reaction with the aminotyrosyl staphylococcal nuclease at pH 5.0 resulted in specific modification. The spectral studies of the two forms of the modified protein shown in Figure 24 and Figure 25 provide an excellent example of the use of 3-nitrotyrosine as a "reporter" group in the protein. Also of interest was the separation of various forms of the modified enzyme by ion-exchange chromatography.

One of the early suggestions for the use of tetranitromethane concerned differentiation between "free" and "buried" tyrosyl residues.[52] Free residues were assumed to be readily accessible to solvent (e.g., on the surface of a globular protein) while buried residues were considered to be in the interior of the molecule (a more "hydrophobic" region). While this may be true in some cases, it is not likely to be a general occurrence. This argument has been addressed by Myers and Glazer.[45] The example of cytochrome c (horse) is worth consideration. Two of the four tyrosyl residues in this protein are converted to 3-nitrotyrosine on reaction with tetranitromethane at pH 8.0 (0.05 M Tris).[65] The two residues modified, Tyr-47 and Tyr-67, are located in the interior while the two residues on the surface of the protein, Tyr-74 and Tyr-97, are not modified. Myers and Glazer emphasize the importance

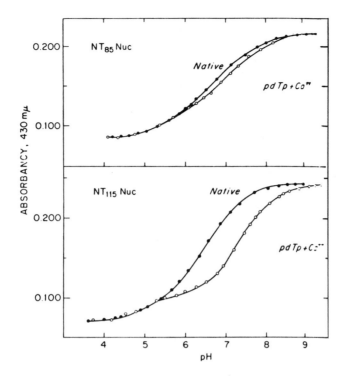

FIGURE 24.   Spectrophotometric titrations of the single nitrotyrosyl hydroxyl group of mononitrotyrosyl 85-Staphylococcal nuclease(NT$_{85}$ Nuc) and mononitrotyrosyl-115-Staphylococcal nuclease(NT$_{115}$ Nuc). The upper panel represents the titration of NT$_{85}$ Nuc in the presence(pdTp + Ca^{++}) and absence (native) of 0.21 mM deoxythymidine 3',5'-di-phosphate(pdTp) and 10 mM CaCl$_2$ while the lower panel represents an identical experiment with NT$_{115}$ Nuc. Solution contained 0.05 mM protein in 0.1 N NaCl. (From Cuatrecasas, P., Fuchs, S., and Anfinsen, C. B., *J. Biol. Chem.*, 243, 4787, 1968. With permission.)

of viewing tetranitromethane (for example) as an organic compound which is more soluble in organic solvents (hydrophobic) than in water. Thus, it is not unreasonable to suggest that reaction at a given residue might, in fact, reflect selective partitioning of the reagent, in this case tetranitromethane, into the microenvironment around the given residue. An analogous situation concerns this increased reaction of phospholipase with tetranitromethane in the presence of substrate (egg yolk lysolecithin).[60] In this case, tetranitromethane is considered to be concentrated in substrate micelles and therefore has an increased rate of reaction with residues involved in substrate binding. This is not to say that accessibility of a residue to the solvent environment is not a factor in modification. Generally, denaturation of a protein (for example see Figure 26) results in increasing modification.

An interesting effect of substrate on the reaction of an enzyme with tetranitromethane has been reported by Christen and Riordan.[66] Generally, protection of an enzyme from loss of activity secondary to chemical modification with the concomitant lack of modification of an amino acid residue is observed in the presence of substrate. Christen and Riordan, however, observed that aspartate aminotransferase was readily inactivated by tetranitro-methane (Figure 27) only in the presence of both substrates, glutamate and α-keto-glutarate ("syncatalytic" modification). This inactivation was associated with the modification of an additional tyrosyl residue. The modified enzyme has some interesting spectral properties (Figures 28 and 29).

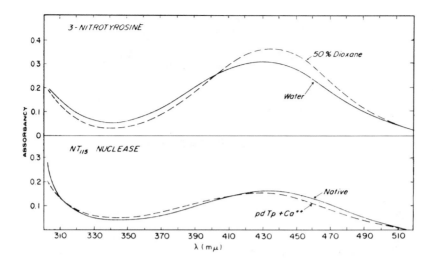

FIGURE 25. Comparison of the absorption spectra of 0.075 m*M* 3-nitrotyrosine in aqueous medium and in 50% dioxane with that of 0.04 m*M* NT_{115} Nuc in the absence (native) and presence(pdTp + Ca^{++}) of 0.21 m*M* deoxythymidine 3′,5′-diphosphate and 10 m*M* $CaCl_2$. The spectra were determined in 0.05 *M* Tris-HCl buffer, pH 8.1. (From Cuatrecasas, P., Fuchs, S., and Anfinsen, C. B., *J. Biol. Chem.*, 243, 4787, 1968. With permission.)

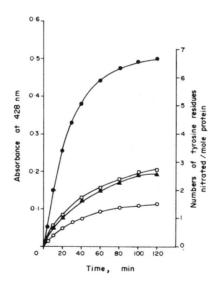

FIGURE 26. The rate of nitration of α_1-antitrypsin with tetranitromethane is shown. The reactions were performed at a protein concentration of 1 mg/mℓ in 0.05 *M* Tris-HCl buffer, pH 8.0, at 25°C. The molar ratio of the tyrosine content to tetranitromethane is ○——○, 1:8.1; ▲——▲, 1:17.2; □——□, 1:34.4; and ●——●, 1:8.6 in 5 *M* guanidine-HCl. (From Busby, T. F. and Gan, J. C., *Int. J. Biochem.*, 6, 835, 1975. With permission.)

Kirschner and Schachman[67] have reported some interesting spectral studies on the catalytic subunit of aspartate transcarbamylase modified by tetranitromethane (0.1 *M* potassium phosphate, pH 6.7). The addition of both substrates, carbamyl phosphate and succinate, resulted in a decrease in absorbance at 430 nm (Figures 30 to 32). This is consistent with the increase

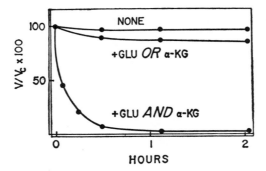

FIGURE 27. Catalysis-dependent inactivation of aspartate aminotransferase by tetranitromethane. Enzyme(21 μ*M*) in 0.05 *M* Tris-Cl(pH 7.5) was incubated at 22°C with 600 μ*M* tetranitromethane in the presence of both 70 m*M* L-glutamate(GLU) and 1.75 m*M* α-ketoglutarate (αKG), in the presence of either glutamate or α-ketoglutarate, and in the absence of either substrate. Tetranitromethane was added at zero time and transaminase activity was determined at the indicated times. The order of addition of enzyme, substrates, and tetranitromethane to the reaction mixture did not influence the course of the reaction. (From Christen, P. and Riordan, J. F., *Biochemistry*, 9, 3025, 1970. With permission.)

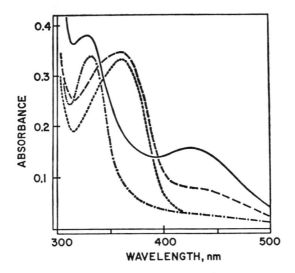

FIGURE 28. UV absorption spectra of aspartate aminotransferase modified with tetranitromethane in the presence and absence of substrates and native pyridoxamine 5′-phosphate and pyridoxal 5′-phosphate enzyme. Conditions for reaction with tetranitromethane were as described in the caption to Figure 27. The enzyme concentration was 18 μ*M* in 0.05 *M* Tris-Cl(pH 8.5); incubated with tetranitromethane in the presence of glutamate and α-ketoglutarate (————), residual activity 3% of control, 1.7 mol of nitrotyrosine/mol enzyme; pyridoxal 5′-phosphate enzyme incubated with tetranitromethane in the absence of substrates (---), residual activity 93% of control, 0.7 mol of nitrotyrosine/mole of enzyme; native pyridoxamine 5′-phosphate enzyme (—·—·); native pyridoxal 5′-phosphate enzyme (···). (From Christen, P. and Riordan, J. F., *Biochemistry*, 9, 3025, 1970. With permission.)

in the pKa of the phenolic hydroxyl (6.25 to 6.62). It should be noted that these investigators found that for the 3-nitrotyrosyl group in the modified aspartate transcarbamylase, $\epsilon_{430} = 4.0 \times 10^3$ cm^{-1} M^{-1} and $\epsilon_{390} = 2.8 \times 10^{-3}$ cm^{-1} M^{-1} (390 nm is the isosbestic point).

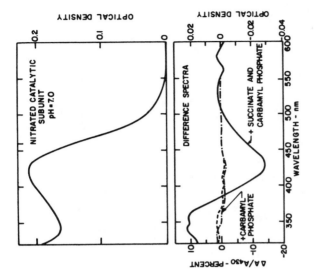

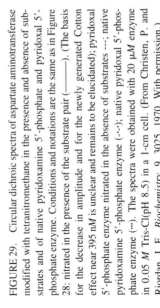

FIGURE 30. UV absorption spectrum and difference spectrum of the catalytic subunit of aspartate transcarbamylase modified with tetranitromethane. The top panel represents the absorption spectrum of nitrated catalytic subunit containing 0.7 nitrotyrosine/polypeptide chain, at a concentration of 3.0 mg/ml in 0.04 M potassium phosphate buffer(pH 7.0). The bottom panel represents the difference spectra of the same sample. The protein *vs.* protein spectrum is represented by (---) which superimposes with the carbamyl phosphate(2 mM) difference spectrum represented by (——·——) above 450 nm. The solid line represents the difference spectrum obtained when succinate(2 mM) and carbamyl phosphate(4 mM) were added to one sample. (From Kirschner, M. W. and Schachman, H. K., *Biochemistry*, 12, 2987, 1973. With permission.)

FIGURE 29. Circular dichroic spectra of aspartate aminotransferase modified with tetranitromethane in the presence and absence of substrates and of native pyridoxamine 5'-phosphate and pyridoxal 5'-phosphate enzyme. Conditions and notations are the same as in Figure 28: nitrated in the presence of the substrate pair (——). (The basis for the decrease in amplitude and for the newly generated Cotton effect near 395 nM is unclear and remains to be elucidated); pyridoxal 5'-phosphate enzyme nitrated in the absence of substrates ---; native pyridoxamine 5'-phosphate enzyme (- -); native pyridoxal 5'-phosphate enzyme (···). The spectra were obtained with 20 μM enzyme in 0.05 M Tris-Cl(pH 8.5) in a 1-cm cell. (From Christen, P. and Riordan, J. F., *Biochemistry*, 9, 3025, 1970. With permission.)

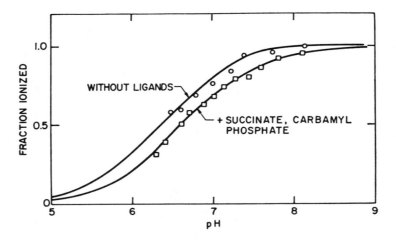

FIGURE 31.   The pH titration of the nitrated catalytic subunit of aspartate trans-carbamylase. The absorbance at 420 nm of the nitrated catalytic subunit was measured in various phosphate buffers in the pH range of 6 to 8. One sample contained no added ligands and the other contained 2 mM succinate and 4 mM carbamyl phosphate. The fraction ionized is plotted vs. pH, so that the pK corresponds to the half-ionization point on this graph. The pK of the unliganded species was found to be 6.25; in the presence of saturating succinate and carbamyl phosphate the pK was 6.62. (From Kirschner, M. W. and Schachman, H. K., *Biochemistry*, 12, 2987, 1973. With permission.)

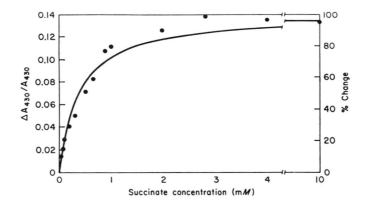

FIGURE 32.   Succinate titration of the nitrated catalytic subunit of aspartate transcarbamylase. The change in absorbance, $\Delta A_{430}/A_{430}$, where A_{430} is the initial absorbance at 430 nm, is given on the ordinate as a function of succinate concentration on the abscissa. All measurements were performed by difference spectroscopy. The nitrated catalytic subunit was at 4 mg/mℓ in 0.04 M potassium phosphate at pH 7.0 containing 2 mM mercaptoethanol, 0.2 M -EDTA, and 4 mM carbamyl phosphate. The solid line represents a theoretical curve for a single class of sites with a dissociation constant of $3.7 \times 10^{-4} M$. (From Kirschner, M. W. and Schachman, H. K., *Biochemistry,* 12, 2987, 1973. With permission.)

There are two other studies on the use of the spectral properties of proteins modified with tetranitromethane. The modification of pancreatic phospholipase[60] has been discussed above. The spectral properties of the modified enzymes have been reported in some detail.[68] Figure

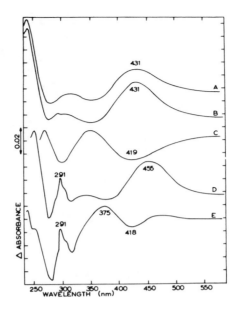

FIGURE 33.   UV difference spectra produced by *N*-acetyl-3-nitrotyrosine ethyl ester and several nitrotyrosine phospholipases upon charge and solvent perturbation. The following are the experimental conditions. Curve A: *N*-acetyl-3-nitrotyrosine ethyl ester(25 μM) in 0.1 *M* NaCl and 50 m*M* sodium acetate, pH 6.0. After equilibration and base-line correction, the sample compartment only was brought to pH 8.0 with solid Tris, and the difference spectrum was recorded. Curve B: horse nitrotyrosine-69 phospholipase A_2(40 μM) treated in the same way as described for curve A. Curve C: *N*-acetyl-3-nitrotyrosine ethyl ester(50 μM) in 0.2 *M* PIPES, pH 6.0; 50% dioxane(V/V) was present in the sample compartment and in the buffer solution of the reference compartment. Curve D: pig nitrotyrosine-69 phospholipase A_2(40 μM) in 0.1 *M* NaCl and 50 m*M* sodium acetate, pH 6.0, in the presence of 5 m*M* *n*-hexadecylphosphocholine. Curve E: horse nitrotyrosine-19 phospholipase A_2(40 μM) treated in the same way as described for curve D. (From Meyer, H., Puijk, W. C., Dijkman, R., Foda-van der Hoorn, M. M. E. L., Pattus, F., Slotboom, A. J., and de Haas, G. H., *Biochemistry*, 18, 3589, 1979. With permission.)

33 shows a difference spectroscopy study of various nitrated derivatives. Curve A is the change in the UV spectrum of *N*-acetyl-3-nitrotyrosine ethyl ester accompanying an increase in pH from 6.0 to 8.0 while curve B shows a similar change in the spectra of equine NO_2Tyr^{69}-phospholipase A_2. Curve C shows the effect of 50% dioxane on the spectrum of the model compound. Curve D shows the effect of *n*-hexadecylphosphocholine on the UV spectrum of porcine NO_2Tyr^{69}-phospholipase A_2 while a similar experiment for equine NO_2Tyr-phospholipase A_2 is presented in curve E. Difference spectra of the dansylamino enzymes (prepared by reaction of dansyl chloride with 3-aminotyrosine obtained by reduction of 3-nitrotyrosine with sodium dithionite[61]) are presented in Figure 34. Difference spectra obtained from the interaction of various phospholipase preparations with *n*-dodecylphosphocholine below critical micellar concentration are shown in Figure 35.

Difference spectra for the interaction of nitrated myosin[69] with calcium ions as a function of pH are shown in Figure 36.

Moravek and co-workers have examined the reaction of human serum albumin with tetranitromethane.[70] This study is of particular interest because 3,5-dinitrotyrosyl residues were reported. While 8 of the 18 tyrosyl residues were readily converted to 3-nitrotyrosine, only 1 or 2 of these residues could be converted to the dinitro derivative.

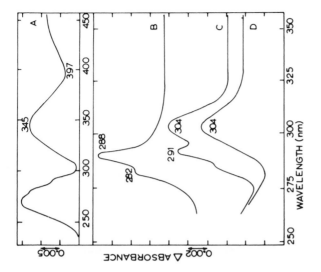

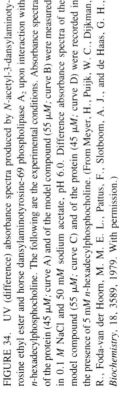

FIGURE 34. UV (difference) absorbance spectra produced by *N*-acetyl-3-dansylaminotyrosine ethyl ester and horse dansylaminotyrosine-69 phospholipase A₂ upon interaction with *n*-hexadecylphosphocholine. The following are the experimental conditions. Absorbance spectra of the protein (45 μM; curve A) and of the model compound (55 μM; curve B) were measured in 0.1 *M* NaCl and 50 m*M* sodium acetate, pH 6.0. Difference absorbance spectra of the model compound (55 μM; curve C) and of the protein (45 μM; curve D) were recorded in the presence of 5 m*M* *n*-hexadecylphosphocholine. (From Meyer, H., Puijk, W. C., Dijkman, R., Foda-van der Hoorn, M. M. E. L., Pattus, F., Slotboom, A. J., and de Haas, G. H., *Biochemistry*, 18, 3589, 1979. With permission.)

FIGURE 35. UV difference absorption spectra produced by the interaction of pig nitrotyrosine-69 (A) phospholipase, (B) pig phospholipase, and (C) pig aminotyrosine-69 phospholipase with *n*-dodecylphosphocholine below the critical micellar concentration. The following are the experimental conditions. Protein concentrations were 40 μ*M* in 0.1 *M* NaCl and 50 m*M* sodium acetate, pH 6.0. Phospholipid was added up to 1.0 m*M*. Solvent perturbation of *N*-acetyl-3-aminotyrosine ethyl ester (curve D) was measured in 0.2 *M* PIPES, pH 6.0; 50% dioxane (V/V) was present in the sample compartment and in the buffer solution of the reference compartment. (From Meyer, H., Puijk, W. C., Dijkman, R., Foda-van der Hoorn, M. M. E. L., Pattus, F., Slotboom, A. J., and de Haas, G. H., *Biochemistry*, 18, 3589, 1979. With permission.)

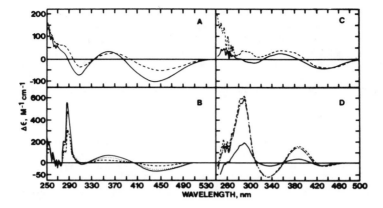

FIGURE 36.   Difference UV absorption spectra for various preparations of myosin modified with tetranitromethane. Shown are absorption difference spectra of nitrated bovine cardiac TN-C(panels A and B), rabbit skeletal TN-C(panel C), and brain modulator protein(panel D) produced by Ca^{2+} addition. The proteins were dissolved in 100 mM MOPS, 50 mM KCl, 1.0 mM EGTA at pH values close to neutrality. A portion of 50 mM $CaCl_2$ was added to the solution in the sample compartment while an equal volume of water was added to the solution in the reference beam. Nitrotyrosine concentrations were approximately 50 to 80 μM corresponding to protein concentrations in the range of 0.6 to 1.0 mg/mℓ. (From McCubbin, W. D., Hincke, M. T., and Kay, C. M., *Can. J. Biochem.*, 57, 15, 1979. With permission.)

REFERENCES

1. **Landsteiner, K.,** *The Specificity of Serological Reactions,* Harvard University Press, Cambridge, 1945.
2. **Fraenkel-Cnrat, H., Bean, R. S., and Lineweaver, H.,** Essential groups for the interaction of ovomucoid (egg white trypsin inhibitor) and trypsin, and for tryptic activity, *J. Biol. Chem.,* 177, 385, 1949.
3. **Riordan, J. F. and Vallee, B. L.,** Diazonium salts as specific reagents and probes of protein conformation, *Meth. Enzymol.,* 25, 521, 1972.
4. **Tabachnick, M. and Sobotka, H.,** Azoproteins. I. Spectrophotometric studies of amino acid azo derivatives, *J. Biol. Chem.,* 234, 1726, 1959.
5. **Tabachnick, M. and Sobotka, H.,** Azoproteins. II. A spectrophotometric study of the coupling of diazotized arsanilic acid with proteins, *J. Biol. Chem.,* 235, 1051, 1960.
6. **Fairclough, G. F., Jr. and Vallee, B. L.,** Arsanilazochymotrypsinogen. The extrinsic Cotton effects of an arsanilazotyrosyl chromophore as a conformation probe of zymogen activation, *Biochemistry,* 10, 2470, 1971.
7. **Gorecki, M., Wilchek, M., and Blumberg, S.,** Modulation of the catalytic properties of α-chymotrypsin by chemical modification at Tyr 146, *Biochem. Biophys. Acta,* 535, 90, 1978.
8. **Robinson, G. W.,** Reaction of a specific tryptophan residue in streptococcal proteinase with 2-hydroxy-5-nitrobenzyl bromide, *J. Biol. Chem.,* 245, 4832, 1970.
9. **Gorecki, M. and Wilchek, M.,** Modification of a specific tyrosine residue of ribonuclease A with a diazonium inhibitor analog, *Biochim. Biophys. Acta,* 532, 81, 1978.
10. **Johansen, J. T., Livingston, D. M., and Vallee, B. L.,** Chemical modification of carboxypeptidase A crystals. Azo coupling with tyrosine-248, *Biochemistry,* 11, 2584, 1972.
11. **Harrison, L. W. and Vallee, B. L.,** Kinetics of substrate and product interactions with arsanilazotyrosine-248 carboxypeptidase A, *Biochemistry,* 17, 4359, 1978.
12. **Cueni, L. and Riordan, J. F.,** Functional tyrosyl residues of carboxypeptidase A. The effect of protein structure on the reactivity of tyrosine-198, *Biochemistry,* 17, 1834, 1978.
13. **Roholt, O. A. and Pressman, D.,** Iodination-isolation of peptides from the active site, *Meth. Enzymol.,* 25, 438, 1972.
14. **Filmer, D. L. and Koshland, D. E., Jr.,** Role of tyrosine residues in chymotrypsin action, *Biochem. Biophys. Res. Commun.,* 17, 189, 1964.

15. **Dube, S. K., Roholt, O. A., and Pressman, D.,** The required integrity of tyrosine in chymotrypsin for the preservation of the catalytic function, *J. Biol. Chem.,* 239, 1809, 1964.

16. **Dube, S. K., Roholt, O., and Pressman, D.,** The chemical nature of the enzyme site of rabbit muscle lactic acid dehydrogenase, *J. Biol. Chem.,* 238, 613, 1963.

17. **Huntley, T. E. and Strittmatter, P.,** The reactivity of the tyrosyl residues of cytochrome b_5, *J. Biol. Chem.,* 247, 4648, 1972.

18. **McGowan, E. B. and Stellwagen, E.,** Reactivity of individual tyrosyl residues of horse heart ferricytochrome c toward iodination, *Biochemistry,* 9, 3047, 1970.

19. **Layne, P. P. and Najjar, V. A.,** Evidence for a tyrosine residue at the active site of phosphoglucomutase and its interaction with vanadate, *Proc. Natl. Acad. Sci. U.S.A.,* 76, 5010, 1979.

20. **Azari, P. R. and Feeney, R. E.,** The resistances of conalbumin and its iron complex to physical and chemical treatments, *Arch. Biochem. Biophys.,* 92, 44, 1961.

21. **Silva, J. S. and Ebner, K. E.,** Protection by substrates and α-lactalbumin against inactivation of galactosyltransferase by iodine monochloride, *J. Biol. Chem.,* 255, 11262, 1980.

22. **Izzo, J. L., Bale, W. F., Izzo, M. J., and Roncone, A.,** High specific activity labeling of insulin with ^{131}I, *J. Biol Chem.,* 239, 3743, 1964.

23. **Linde, S., Sonne, O., Hansen, B., and Gliemann, J.,** Monoiodoinsulin labelled in tyrosine residue 16 or 26 of the insulin B-chain. Preparation and characterization of some binding properties, *Hoppe-Seyler's Z. Physiol. Chem.,* 362, 573, 1981.

24. **Hunter, W. M. and Greenwood, F. C.,** Preparation of iodine-131 labelled human growth hormone of high specific activity, *Nature (London),* 194, 495, 1962.

25. **Heber, D., Odell, W. D., Schedewie, H., and Wolfsen, A. R.,** Improved iodination of peptides for radioimmunoassay and membrane radioreceptor assay, *Clin. Chem.,* 24, 796, 1978.

26. **Kurihara, K., Horinishi, H., and Shibata, K.,** Reaction of cyanuric halides with proteins. I. Bound tyrosine residues of insulin and lysozyme as identified with cyanuric fluoride, *Biochim. Biophys. Acta,* 74, 678, 1963.

27. **Gorbunoff, M. J.,** Cyanuration, *Meth. Enzymol.,* 25, 506, 1972.

28. **Gorbunoff, M. J. and Timasheff, S. N.,** The role of tyrosines in elastase, *Arch. Biochem. Biophys.,* 152, 413, 1972.

29. **Coffe, G. and Pudles, J.,** Chemical reactivity of the tyrosyl residues in yeast hexokinase. Properties of the nitroenzyme, *Biochim. Biophys. Acta,* 484, 322, 1977.

30. **Ferguson, S. J., Lloyd, W. J., Lyons, M. H., and Radda, G. K.,** The mitochondrial ATPase. Evidence for a single essential tyrosine residue, *Eur. J. Biochem.,* 54, 117, 1975.

31. **Ferguson, S. J., Lloyd, W. J., and Radda, G. K.,** The mitochondrial ATPase. Selective modification of a nitrogen residue in the β subunit, *Eur. J. Biochem.,* 54, 127, 1975.

32. **Andrews, W. W. and Allison, W. S.,** 1-Fluoro-2,4,-dinitrobenzene modifies a tyrosine residue when it inactivates the bovine mitochondrial F_1-ATPase, *Biochem. Biophys. Res. Commun.,* 99, 813, 1981.

33. **Murachi, T.,** A general reaction of diisopropylphosphorofluoridate with proteins without direct effect on enzymic activities, *Biochim. Biophys. Acta,* 71, 239, 1963.

34. **Murachi, T., Inagami, T., and Yasui, M.,** Evidence for alkylphosphorylation of tyrosyl residues of stem bromelain by diisopropylphosphorofluoridate, *Biochemistry,* 4, 2815, 1965.

35. **Inagami, T. and Murachi, T.,** Reaction of diisopropylphosphorofluoridate with *N*-acetyl-L-tyrosinamide, *J. Biochem. (Tokyo),* 68, 419, 1970.

36. **Means, G. E. and Wu, H.-L.,** The reactive tyrosine residue of human serum albumin: characterization of its reaction with diisopropylfluorophosphate, *Arch. Biochem. Biophys.,* 194, 526, 1979.

37. **Wieland, T. and Schneider, G.,** *N*-acylimidazoles as acyl derivatives of high energy, *Ann. Chem. Justus Liebigs,* 580, 159, 1953.

38. **Stadtman, E. R. and White, F. H., Jr.,** The enzymic synthesis of *N*-acetylimidazole, *J. Am. Chem. Soc.,* 75, 2022, 1953.

39. **Stadtman, E. R.,** On the energy-rich nature of acetyl imidazole, an enzymatically active compound, in *A Symposium on the Mechanism of Enzyme Action,* McElroy, W. D. and Glass, B., Eds., Johns Hopkins Press, Baltimore, 1954, 581.

40. **Riordan, J. F. and Vallee, B. L.,** Acetylation, *Meth. Enzymol.,* 11, 565, 1967.

41. **Karibian, D., Jones, C., Gertler, A., Dorrington, K. J., and Hofmann, T.,** On the reaction of acetic and maleic anhydrides with elastase. Evidence for a role of the NH_2-terminal valine, *Biochemistry,* 13, 2891, 1974.

42. **Ohnishi, M., Suganuma, T., and Hiromi, K.,** The role of a tyrosine residue of bacterial liquefying α-amylase in the enzymatic hydrolysis of linear substrates as studied by chemical modification with acetic anhydride, *J. Biochem. (Tokyo),* 76, 7, 1974.

43. **Simpson, R. T., Riordan, J. F., and Vallee, B. L.,** Functional tyrosyl residues in the active center of bovine pancreatic carboxypeptidase A, *Biochemistry,* 2, 616, 1963.

44. **Riordan, J. F., Wacker, W. E. C., and Vallee, B. L.,** *N*-Acetylimidazole: a reagent for determination of "free" tyrosyl residues of proteins, *Biochemistry,* 4, 1758, 1965.
45. **Myers, B. II and Glazer, A. N.,** Spectroscopic studies of the exposure of tyrosine residues in proteins with special reference to the subtilisins, *J. Biol. Chem.,* 246, 412, 1971.
46. **Riordan, J. F. and Vallee, B. L.,** *O*-Acetyltyrosine, *Meth. Enzymol.,* 11, 570, 1967.
47. **Riordan, J. F. and Vallee, B. L.,** *O*-Acetyltyrosine, *Meth. Enzymol.,* 25, 500, 1972.
48. **Tildon, J. T. and Ogilvie, J. W.,** The esterase activity of bovine mercaptalbumin. The reaction of the protein with *p*-nitrophenyl acetate, *J. Biol. Chem.,* 247, 1265, 1972.
49. **Herriott, R. M.,** Reactions of native proteins with chemical reagents, *Adv. Protein Chem.,* 3, 169, 1947.
50. **Astrup, T.,** Inactivation of thrombin by means of tetranitromethane, *Acta Chem. Scand.,* 1, 744, 1948.
51. **Riordan, J. F., Sokolovsky, M., and Vallee, B. L.,** Tetranitromethane. A reagent for the nitration of tyrosine and tyrosyl residues in proteins, *J. Am. Chem. Soc.,* 88, 4104, 1966.
52. **Sokolovsky, M., Riordan, J. F., and Vallee, B. L.,** Tetranitromethane. A reagent for the nitration of tyrosyl residues in proteins, *Biochemistry,* 5, 3582, 1966.
53. **Sokolovsky, M., Harell, D., and Riordan, J. F.,** Reaction of tetranitromethane with sulfhydryl groups in proteins, *Biochemistry,* 8, 4740, 1969.
54. **Riordan, J. F. and Vallee, B. L.,** Nitration with tetranitromethane, *Meth. Enzymol.,* 25, 515, 1972.
55. **Cuatrecasas, P., Fuchs, S., and Anfinsen, C. B.,** The tyrosyl residues at the active site of staphylococcal nuclease. Modifications by tetranitromethane, *J. Biol. Chem.,* 243, 4787, 1968.
56. **Lane, R. S. and Dekker, E. E.,** Oxidation of sulfhydryl groups of bovine liver 2-keto-4-hydroxyglutarate aldolase by tetranitromethane, *Biochemistry,* 11, 3295, 1972.
57. **Hugli, T. E. and Stein, W. H.,** Involvement of a tyrosine residue in the activity of bovine pancreatic deoxyribonuclease A, *J. Biol. Chem.,* 246, 7191, 1971.
58. **Rickert, W. S. and McBride-Warren, P. A.,** Structural and functional determinants of *Mucor miehei* protease. IV. Nitration and spectrophotometric titration of tyrosine residues, *Biochim, Biophys. Acta,* 371, 368, 1974.
59. **Busby, T. F. and Gan, J. C.,** The reaction of tetranitromethane with human plasma α_1-antitrypsin, *Int. J. Biochem.,* 6, 835, 1975.
60. **Meyer, H., Verhoef, H., Hendriks, F. F. A., Slotboom, A. J., and de Haas, G. H.,** Comparative studies of tyrosine modification in pancreatic phospholipases. I. Reaction of tetranitromethane with pig, horse and ox phospholipases A_2 and their zymogens, *Biochemistry,* 18, 3582, 1979.
61. **Sokolovsky, M., Riordan, J. F., and Vallee, B. L.,** Conversion of 3-nitrotyrosine to 3-aminotyrosine in peptides and proteins, *Biochem. Biophys. Res. Commun.,* 27, 20, 1967.
62. **Riordan, J. F., Sokolovsky, M., and Vallee, B. L.,** Environmentally sensitive tyrosyl residues. Nitration with tetranitromethane, *Biochemistry,* 6, 358, 1967.
63. **Riordan, J. F., Sokolovsky, M., and Vallee, B. L.,** The functional tyrosyl residues of carboxypeptidase A. Nitration with tetranitromethane, *Biochemistry,* 6, 3609, 1967.
64. **Muszynska, G. and Riordan, J. F.,** Chemical modification of carboxypeptidase A crystals. Nitration of tyrosine-248, *Biochemistry,* 15, 46, 1976.
65. **Skov, K., Hofmann, T., and Williams, G. R.,** The nitration of cytochrome c, *Can. J. Biochem.,* 47, 750, 1969.
66. **Christen, P. and Riordan, J. F.,** Syncatalytic modification of a functional tyrosyl residue in aspartate aminotransferase, *Biochemistry,* 9, 3025, 1970.
67. **Kirschner, M. W. and Schachman, H. K.,** Conformational studies on the nitrated catalytic subunit of aspartate transcarbamylase, *Biochemistry,* 12, 2987, 1973.
68. **Meyer, H., Puijk, W. C., Dijkman, R., Foda-van der Hoorn, M. M. E. L., Pattus, F., Slotboom, A. J., and de Haas, G. H.,** Comparative studies of tyrosine modification in pancreatic phospholipases. II. Properties of the nitrotyrosyl, aminotyrosyl, and dansylaminotyrosyl derivatives of pig, horse, and ox phospholipases A_2 and their zymogens, *Biochemistry,* 18, 3589, 1979.
69. **McCubbin, W. D., Hincke, M. T., and Kay, C. M.,** The utility of the nitrotyrosine chromophore as a spectroscopic probe in troponin C and modulator protein, *Can. J. Biochem.,* 57, 15, 1979.
70. **Moravek, L., Saber, M. A., and Meloun, B.,** Steric accessibility of tyrosine residues in human serum albumin, *Collect Czech. Chem. Commun.,* 44, 1657, 1979.

Chapter 4

THE MODIFICATION OF CARBOXYL GROUPS

Although the selective modification of carboxyl groups was the subject of one of the early attempts[1] to adapt organic chemistry to the systematic study of the relationship between structure and function in proteins, serious study in the area did not develop at a significant rate until the application of water soluble carbodiimide–mediated attack by a protein-bound carboxyl group on a suitable nucleophile (Figure 1) (i.e., radiolabeled glycine methyl ester, norleucine methyl ester) by Koshland and Hoare.[2]

The above observation that the majority of studies on the chemical modification of carboxyl groups utilize the carbodiimide-mediated reaction(s) is not intended to indicate that there are not other approaches to the modification of aspartyl and glutamyl residues in proteins. Indeed, there are examples of carboxyl group modification with reagents expected to react far more effectively with other nucleophiles. An example of this is the reaction of iodoacetamide with ribonuclease T₁ to form the glycolic acid derivative of the glutamic acid residue as elegantly shown by Takahashi and co-workers.[3] Another example is the modification of a specific carboxyl group in pepsin by *p*-bromophenacyl bromide[3] (the use of *p*-bromophenacyl bromide in the specific modification of proteins is not uncommon but is generally associated with the modification of cysteine, histidine, or methionine). In the study of pepsin, optimal inactivation (approximately 12-fold molar excess of reagent, 3 hr, 25°C) was obtained in the pH range of 1.5 to 4.0 with a rapid decrease in the extent of inactivation at pH 4.5 and above (the effect of pH greater than 5.5 to 6.0 on the modification of pepsin cannot be studied because of irreversible denaturation of pepsin at pH 6.0 and above). In studies with a 10% molar excess of *p*-bromophenacyl bromide at pH 2.8, 37°C, 3 hr, complete inactivation of the enzyme was obtained concomitant with the incorporation of 0.93 mol of reagent/mole of pepsin (assessed by bromide analysis). Attempts to reactivate the modified enzyme with a potent nucleophile such as hydroxylamine were unsuccessful but reactivation could be obtained with sulfhydryl-containing reagents (i.e., β-mercaptoethanol, 2,3-dimercaptopropanol, thiophenol). It has been subsequently established that reaction occurs at the β-carboxy group of an aspartic acid residue (formation of 2-*p*-bromophenyl-1-ethyl-2-one β-aspartate).[5] These investigators noted that reduction of enzyme under somewhat harsh conditions (LiBH₄ in tetrahydrofuran) resulted in the formation of homoserine.

Active-site directed reactions with reactive epoxy functional groups have proved useful in several studies of the role of carboxyl groups in proteins. Tang and co-workers have used 1,2-epoxy-3-(*p*-nitrophenoxy) propane for the modification of catalytically important carboxyl groups in pepsin.[6,7] Active-site directed inactivation of lysozyme with an epoxy function (2′,3′-epoxypropyl β-glycoside of di-(*N*-acetyl-D-glucosamine)) which reacts with the β-carboxyl group of aspartic acid has been described.[8,9]

Diazo compounds have proved useful for some time in the esterification of the carboxyl groups of proteins. This is particularly true of diazomethane. The use of this compound was reviewed 15 years ago by the late Philip Wilcox,[10,11] and we are not aware of the extensive use of this compound during the past decade. Various α-keto diazo derivatives have proved particularly fruitful in the study of acid proteinases. Rajagopalan, Stein, and Moore[12] demonstrated that pepsin was inactivated by diazoacetyl-L-norleucine methyl ester. During the course of these studies, it was observed that cupric ions greatly enhanced both the rate and specificity of the modification. Originally it was suggested that cupric ions blocked nonspecific reaction with carboxyl groups not at the active site. Subsequently it was shown that cupric ions and diazoacetyl-norleucine methyl ester formed a highly reactive species, presumably a copper-complexed carbene, which then reacted with a specific *protonated* carboxyl group at the active site of pepsin.[13,14] The modification of carboxyl groups in a variety of

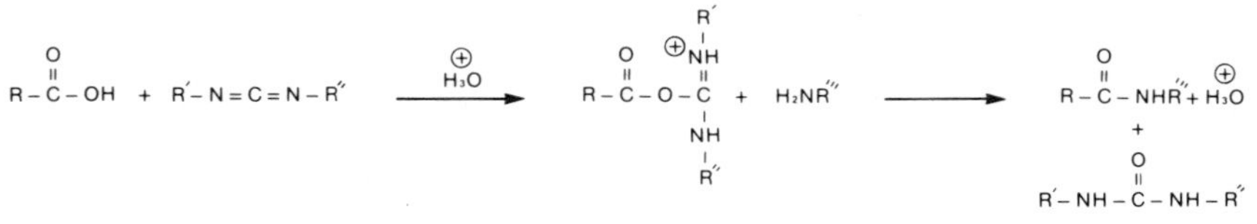

FIGURE 1.   The reaction of a protein-bound carboxyl group and a nucleophile mediated by a carbodiimide.

acid proteinases with a variety of α-keto diazo compounds is shown in Table 1. These diazo compounds are by no means specific for carboxyl group modification in protein. Benzyl-oxycarbonyl-phenylalanyldiazomethylketone has been shown to modify cathepsin B_1, presumably by reaction with the active site sulfhydryl group.[15] Other possible side reactions of α-keto diazo compounds have been reviewed by Widner and Viswanatha.[16] These side reactions result primarily from the oxidative modification of tryptophan, methionine, tyrosine, and cystine. These side reactions can be virtually obviated by vigorous exclusion of oxygen from the reaction and the addition of an oxygen scavenger (e.g., $Na_2S_2O_4$).

Other approaches to the conversion of carboxylic acid functional groups to methyl or ethyl esters have been considered. Trialkyloxonium fluoroborate salts (Figure 2) have proved effective. Raftery and co-workers used triethyloxonium fluoroborate to modify the β-carboxyl groups of an aspartic residue essential for the enzymatic activity of lysoszyme.[17,18] Paterson and Knowles[19] used trimethyloxonium fluoroborate to determine the number of carboxyl groups in pepsin which are essential for catalytic activity. This article discusses in some depth the rigorous precautions necessary for the preparation of this reagent. This reagent is highly reactive and considerable care is required for its introduction into the reaction mixture containing protein. The reaction is performed at pH 5.0 (0.020 M sodium citrate, pH maintained at 5.0 with 2.5 M NaOH). These investigators also report the preparation of the [14C]-labeled reagent from sodium methoxide and [14C] methyliodide.

The enzymatic methylation of protein carboxyl groups has been reported.[20]

Woodward and co-workers[21,22] developed N-ethyl-5-phenylisoxazolium-3'-sulfonate (Woodward's reagent K) and various other N-alkyl-5-phenylisoxazolium fluoroborates as reagents for the "activation" of carboxyl groups for synthetic purposes. Shaw and co-workers[23] used N-ethyl-5-phenylisoxazolium-3-sulfonate, the N-methyl and N-ethyl derivatives of 5-phenylisoxazolium fluoroborate or N-methylbenzisoxazolium fluoroborate (Figure 3) to activate carboxy groups on trypsin for subsequent modification with methylamine or ethylamine. The extent of modification obtained ranged from approximately 3 residues modified (N-methyl-5-phenylisoxazolium fluoroborate or N-ethyl-5-phenylisoxazolium fluoroborate, pH 3.80, 20°C, 80 min) to approximately 11 residues modified (N-methyl-5-phenylisoxazolium fluoroborate, pH 6.0, 20°C, 10 min. Reagent decomposition occurs quite rapidly, even at ice-bath temperature (2°C). The modification appears fairly selective for carboxyl groups although some modification of lysine was observed under conditions where extensive modification was obtained (250-fold molar excess of N-methyl-5-phenylisoxazolium fluoroborate, pH 4.75, 72 min, 20°C, methylamine as the attacking nucleophile). Saini and Van Etten[24] reported on the reaction of N-ethyl-5-phenylisoxazolium-3'-sulfonate with human prostatic acid phosphatase. The modification was performed with a 4,000- to 10,000-fold molar excess of reagent in 0.020 M pyridinesulfonic acid, pH 3.6 at 25°C. Ethylamine was utilized as the attacking nucleophile to determine the extent of modification. A substantial number of carboxyl groups in the protein were modified under these experimental conditions. Arana and Vallejos[25] have compared the reaction of chloroplast coupling factor with N-ethyl-5-phenylisoxazolium-3'-sulfonate (Woodward's Reagent K) and dicyclohexylcarbodiimide. Reaction with Woodward's Reagent K was accomplished at 25°C in 0.040 M-Tricine, pH

Table 1
REACTION OF DIAZOACETYL COMPONDS WITH ACID PROTEINASES

Reagent	Solvent	Enzyme	Ref.
Diazoacetylnorleucine methyl ester	0.04 *M* sodium acetate, 0.1 *M* Cu (OAc)$_2$, pH 5.0	Pepsin	1
Tosyl-L-phenylalanyldiazomethane[a]	6.25 m*M* acetate, 1 m*M* CuCl$_2$, pH 5.4	Pepsin	2
α-Diazo-*p*-bromoacetophenone	6 m*M* acetate, 2 m*M* CuCl$_2$, pH 5.0	Pepsin	3
1-Diazo-3-(2,4-dinitrophenylamido)-propanone	0.1 *M* acetate, 1 m*M* CuSO$_4$, pH 5.6	Pepsin	4
Diazoacetyl-L-phenylalanine methyl ester	0.05 *M* sodium acetate, 2 m*M* Cu(OAc)$_2$, pH 5.0	Pepsin	5
Diazoacetylglycine[b] methyl ester	0.02 *M* sodium acetate, 1 m*M* Cu(OAc)$_2$, pH 5.6	Pepsin	6
Diazoacetic acid[b] methyl ester	0.02 *M* sodium acetate, 1 m*M* Cu(OAc)$_2$, pH 5.6	Pepsin	6
1-Diazo-4-phenyl-2-butanone	0.04 *M* sodium acetate, 0.1 m*M* CuSO$_4$, pH 5.5	Pepsin	7
N-Diazoacetyl-*N'*-2,4-dinitrophenylethylenediamine	0.04 *M* sodium acetate, 1 m*M* Cu(OAc)$_2$, pH 5.5	Awamorin[c]	8
Diazoacetylnorleucine methyl ester	0.02 *M* sodium acetate, Cu(OAc)$_2$[d], pH 5.0—5.6	*Rhizopus chinensis* acid proteinase	9
		Aspergillus saitoi acid proteinase	9
		Mucor pusillus acid proteinase	9
		Calf rennin	9
4-(3-diazo-2-oxopropylidene)-[e] 2,2,6,6-tetramethylpiperidine-1-oxyl	0.1 *M* sodium acetate, 0.05 m*M* Cu(OAc)$_2$, pH 5.5	Pepsin	10

[a] L-1-diazo-4-phenyl-3-tosylamidobutanone.
[b] Both of these compounds inhibited pepsin in the presence of cupric ions but there was a greater extent of carboxyl group modification than seen with diazoacetylnorleucine methyl ester.
[c] An acid proteinase isolated from the mold *Aspergillus awamori*.
[d] 40– to 220-fold molar excess of cupric ions (as acetate salt) with respect to the individual enzyme is used with a 40- to 45-fold molar excess of diazoacetylnorleucine methyl ester.
[e] Other spin-labeled compounds were prepared in this study including the cis and trans isomers of 3-(4-diazo-3-oxo-1-butenyl)-2,2,5,5-tetramethylpyrroline-1-oxyl.

REFERENCES FOR TABLE 1

1. **Rajagopalan, T. G., Stein, W. H., and Moore, S.,** The inactivation of pepsin by diazoacetylnorleucine methyl ester, *J. Biol. Chem.,* 241, 4295, 1966.
2. **Delpierre, G. R. and Fruton, J. S.,** Specific inactivation of pepsin by a diazo ketone, *Proc. Natl. Acad. Sci. U.S.A.,* 56, 1817, 1966.
3. **Erlanger, B. F., Vratsanos, S. M., Wassermann, N., and Cooper, A. G.,** Stereochemical investigation of the active center of pepsin using a new inactivator, *Biochem. Biophys. Res. Commun.,* 28, 203, 1967.
4. **Kozlov, L. V., Ginodman, L. M., and Orekhovich, V. N.,** Inactivation of pepsin with aliphatic diazocarbonyl compounds, *Biokhimiya,* 32, 1011, 1967.
5. **Bayliss, R. S., Knowles, J. R., and Wybrandt, G. B.,** An aspartic acid residue at the active site of pepsin. The isolation and sequence of the heptapeptide, *Biochem. J.,* 113, 377, 1969.
6. **Lundblad, R. L. and Stein, W. H.,** On the reaction of diazoacetyl compounds with pepsin, *J. Biol. Chem.,* 244, 154, 1969.
7. **Fry, K. T., Kim, O.-K., Spona, J., and Hamilton, G. A.,** Site of reaction of a specific diazo inactivator of pepsin, *Biochemistry,* 9, 4624, 1970.
8. **Kovaleva, G. G., Shimanskaya, M. P., and Stepanov, V. M.,** The site of diazoacetyl inhibitor attachment to acid proteinase of *Aspergillus awamori* — An analog of penicillopepsin and pepsin, *Biochem. Biophys. Res. Commun.,* 49, 1075, 1972.

Table 1 (continued)

9. **Takahashi, K., Mizobe, F., and Chang, W.-J.,** Inactivation of acid proteases from *Rhizopus chinensis, Aspergillus saitoi* and *Mucor pusillus* and calf rennin by diazoacetyl norleucine methyl ester, *J. Biochem.,* 71, 161, 1972.
10. **Nakayama, S.-I., Nagashima, Y., Hoshino, M., Moriyama, A., Takahashi, K., Uematsu, Y., Watanabe, T., and Yoshida, M.,** Spin-labelling of porcine pepsin and *Rhizopus chinensis* acid protease by diazoketone reagents, *Biochem. Biophys. Res. Commun.,* 101, 658, 1981.

Structure	Nomenclature	Abbreviation
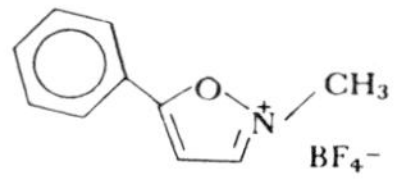	*N*-Methyl-5-phenylisoxazolium fluoroborate	MPI
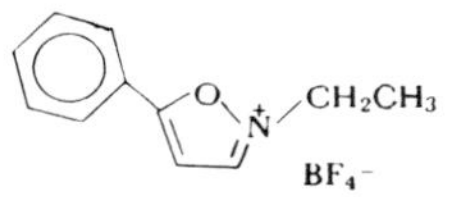	*N*-Ethyl-5-phenylisoxazolium fluoroborate	EPI
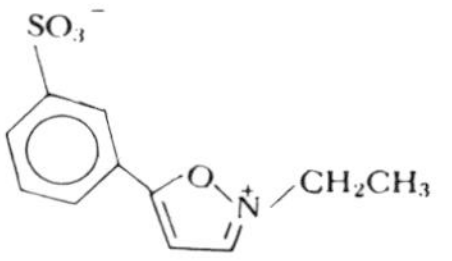	*N*-Ethyl-5-phenylisoxazolium-3'-sulfonate Woodwards K reagent)	K
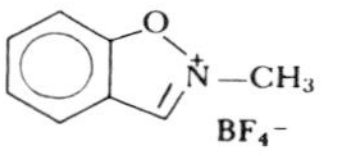	*N*-Methylbenzisoxazolium fluoroborate	MBI

FIGURE 2.   The structures of some isoxazolium salts. (From Bodlaender, P., Feinstein, G., and Shaw, E., *Biochemistry,* 8, 4941, 1969. With permission.)

7.9 while reaction with dicyclohexylcarbodiimide was accomplished at 30°C in 0.040 M MOPS, pH 7.4. ATP and derivatives such as ADP and inorganic phosphate protect against the loss of activity occurring upon reaction with Woodward's Reagent K but do not have any effect on inactivation by dicyclohexylcarbodiimide. The reverse was seen with divalent cations such as Ca^{++}. The modification of an essential carboxyl group in pancreatic phospholipase A_2 by 5-ethyl-5-phenylisoxazolium-3-sulfonate has been reported.[26] The reaction was performed in 0.01 M sodium phosphate, pH 4.75 (pH stat) at 25°C. A second-order rate constant of $k_2 = 25.5$ $M^{-1}min^{-1}$ was obtained for the loss of catalytic activity. This rate inactivation is increased more than twofold in the presence of 30 mM $CaCl_2$ (69.3 $M^{-1}min^{-1}$). Quantitative information on the extent of modification is obtained with [^{14}C] glycine ethyl ester. It is of interest that treatment with a water-soluble carbodiimide, 1-(3'-dimethyl-aminopropyl)-3-ethylcarbodiimide, results in the loss of catalytic activity in a reaction with characteristics different from those seen with Woodward's Reagent K.

The use of carbodiimide-mediated modification of carboxyl functional groups in proteins is by far the most widely used method for the study of such functional groups. The most popular approach utilizes a water-soluble carbodiimide as the activating agent as introduced by Hoare and Koshland.[2] These investigators show that virtually quantitative modification

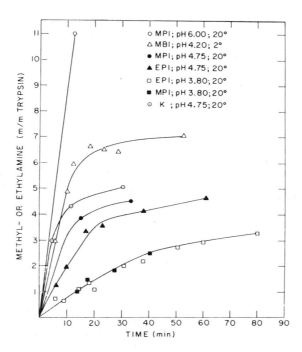

FIGURE 3. The reaction of isoxazolium salts with β-trypsin at various pH values. The isoxazolium salts were MPI, *N*-methyl-5-phenylisoxazolium fluoroborate; EPI, *N*-ethyl-5-phenylisoxazolium fluoroborate; K, *N*-ethyl-5-phenylisoxazolium-3'-sulfonate (Woodwards K reagent); and MBI, *N*-methyl-benzisoxazolium fluoroborate. In all reactions, the initial reagent and protein concentrations were 10 and 5 mg/mℓ, respectively, except for the reaction with K reagent in which the respective concentrations were 5 and 1.5 mg/mℓ. Each point was obtained by removing a portion from the reaction mixture and adjusting its pH to 2.5 with formic acid at the indicated times. Amino acid analyses were performed on the gel-filtered samples to determine methyl- or ethylamine content per mole of trypsin. (From Bodlaender, P., Feinstein, G., and Shaw, E., *Biochemistry*, 8, 4941, 1969. With permission.)

of the carboxyl groups in lysozyme (8.1/11), chymotrypsin (15.5/17), and trypsin (12.5/11) occurred with 0.1 *M* carbodiimide at pH 4.75 with 1.0 *M* glycine methyl ester in a pH stat at 25°C. The possibility of side reaction was discussed with reference to the possible modification of the phenolic hydroxyl of tyrosine to form the *O*-arylisourea. Modification of the active site serine residue in α-chymotrypsin with 1-cyclohexyl-3-(2-morpholinyl-(4)-ethyl) carbodiimide metho-*p*-toluenesulfonate at neutral pH (Figure 4) has been reported.[27] Note in particular the increased rate at neutral or alkaline pH. This is not consistent with carboxyl group modification which *requires a protonated carboxyl group*. The modification of the active site cysteinyl group in papain has been reported[28] as occurring under conditions (pH 4.75, 25°C) where 6/14 carboxyl groups are modified together with 9/19 tyrosyl residues. Modification of the tyrosyl residues is reversed by 0.5 *M* hydroxylamine, pH 7.0 (5 hr at 25°C) as first demonstrated by Carraway and Koshland.[29] Despite the problems with side reaction, modification of carboxyl group in proteins with a water-soluble carbodiimide and an appropriate nucleophile (e.g., [^{14}C] glycine ethyl ester, norleucine methyl ester — easily detected by amino acid analysis, aminomethylsulfonic acid) has proved extremely useful and is reviewed in several articles.[30] It should be noted that ammonium ions can be used as the attacking nucleophile to generate asparaginyl and glutaminyl residues from "exposed" carboxyl groups. The modification was accomplished in 5.5 *M* NH$_4$Cl at pH 4.75 for 3 hr at 25°C. Under these conditions, approximately 11 of the 15 free carboxyl groups in chy-

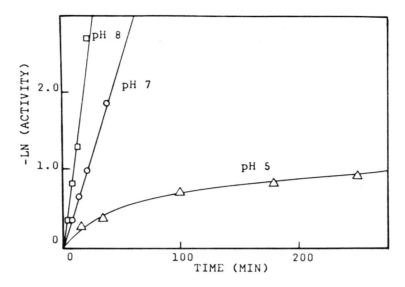

FIGURE 4.   The effect of pH on the inactivation of α-chymotrypsin by 1-cyclohexyl-3-[2-morpholinyl-(4)-ethyl]carbodiimide metho-*p*-toluenesulfonate(CMC). The experiments were performed in 0.1 *M* KCl with α-chymotrypsin(0.1 mg/mℓ) and 10 m*M* CMC at 25°C. pH was maintained with a pH stat. (From Banks, T. E., Blossey, B. F., and Schafer, J. A., *J. Biol. Chem.*, 244, 6323, 1969. With permission.)

motrypsinogen were converted to the corresponding amide.[31] 1,2-Diaminoethane or diaminomethane can be coupled to aspartic acid residues to produce a trypsin-sensitive bond.[32] Examples of the use of this approach are given in Table 2.

In addition to the information presented in Table 2, certain examples of carbodiimide-mediated modification should be discussed in greater detail. The inactivation of yeast hexokinase[33] by reaction with 1-cyclohexyl-3-(2-morpholinoethyl) carbodiimide metho-*p*-toluenesulfonate is shown in Figure 5. These experiments should be compared to those shown in Figure 6 with the same reagent in the presence of nitrotyrosine ethyl ester. The use of the nitrotyrosine derivative as the attacking nucleophile allowed the introduction of a "reporter group" onto the glutamyl residue modified in this study. The use of the nitro-tyrosyl derivative also permitted the facile determination of the extent of modification (in 0.1 *M* NaOH, thus the nitrotyrosyl derivative had A_{max} at 430 nm ($\epsilon = 4.6 \times 10^{-3}$ $M^{-1}cm^{-1}$). Studies on the use of this carbodiimide in the modification of phosphorylase b have been reported.[34] Figure 7 shows the reaction of rabbit muscle phosphorylase with 1-cyclohexyl-3-(2-morpholinyl-(4)-ethyl) carbodiimide metho-*p*-toluenesulfonate (CMC) in the absence of added nucleophile while Figure 8 shows the effect of added nucleophiles on the rate of inactivation.

There are several observations on the use of 1-ethyl-3-(3-dimethylaminopropyl)-carbodiimide (EDC). Figure 9 compares the rate of inactivation of yeast enolase[35] with this reagent and several other carbodimides. An exogenous nucleophile was not used in these experiments. Note that 1-ethyl-3-(4-azonia-4,4-dimethylpentyl)-carbodiimide iodide (EAC) appears to be far more effective than the other two carbodiimides under these reaction conditions. The selective modification of a single aspartyl residue at the active site of lysozyme (Asp-101) was accomplished by the use of a low molar excess (five to tenfold) of carbodiimide. A variety of attacking nucleophiles were used in this study (the modification reactions were performed at pH 5.0 maintained with HCl during the reaction). Of particular interest is the success achieved in the separation of the products of the reaction by ion-exchange chro-

Table 2

THE MODIFICATION OF CARBOXYL GROUPS IN PROTEINS

Protein	Reaction conditions	Carbodiimide	Molar excess	Nucleophile	Carboxyl groups modified	Other functional groups modified	Ref.
Lysozyme	pH 4.75, 25°C, pH stat,[a] 0.1 M carbodiimide	N-benzyl-N'-3-[b] dimethylamino-propylcarbodiimide	140	1.0 M glycine methyl ester	2.1/11 (5 min) 4.7/11 (60 min) 8.1/11 (5—6 hr)	— — —	1
Lysozyme			140	0.1 M nitro-tyrosine ethyl ester	1.4/11[c]	—	1
Trypsin			250	1.0 M glycine methyl ester	4.6/11 (5 min) 8.8/11 (60 min) 12.5/11 (5—6 hr)[d]	— — —	1
Chymotrypsin			250	1.0 M glycine methyl ester	6.2/17 (5 min) 11.8/17 (60 min) 15.5/17 (5—6 hr)	— — —	1
Trypsinogen	pH 4.5, 25°C	1-ethyl-3-[e] (3-dimethylamino-propyl) carbodiimide EDC	25	1.0 M glycine[f] ethyl ester	—	—	2
Chymo-trypsinogen	pH 4.0, 25°C, pH stat; pH 6.0, 25°C, pH stat; pH 8.0, 25°C, pH stat	EDC		1.0 M glycine ethyl ester	13/14[g] 10/14 3/14	— —	3 3 3
Chymotrypsin	pH 4.75, 25°C, pH stat	EDC		1.0 M glycine methyl ester	12.7[h] 15.6[i] 10.6 13.5[i]	—	4 4
Lysozyme	pH 4.75, 25°C, pH stat	BDC		0.25 M amino-methanesulfonic acid; 1.0 M glycine methyl ester	8.5—9.5[j] 6.5—7.5[j]	—	5 5

Table 2 (continued)
THE MODIFICATION OF CARBOXYL GROUPS IN PROTEINS

Protein	Reaction conditions	Carbodiimide	Molar excess	Nucleophile	Carboxyl groups modified	Other functional groups modified	Ref.
Trypsin	pH 4.75, 25°C, pH stat	EDC		1.0 M glycinamide	7.9[k], 1.7[k]	3—5 Tyr[l]	6
Albumin	pH 4.75	EDC		Glycine methyl ester or L-argininamide	1[m]	—	7
Chymotrypsin	pH 4.0[n]	EDC		Glycine ethyl ester	15/15	—	8
Papain	pH 4.75, 25°C	EDC	300—1200	Glycine ethyl ester	6/14	6—10 Tyr,[o] active site cysteinyl residue	9
Yeast hexokinase	0.1 M phosphate, pH 6.0, 20°C[p]	1-cyclohexyl-3-(2-morpholino-ethyl)carbodiimide metho-p-toluenesulfonate[q]	500—3000	Nitrotyrosine[r] ethyl ester	1	Cysteinyl (2)	10
Phosphorylase[b]	pH 5.1, 25°C	CMC		Glycine[s] ethyl ester	3[t]	—	11
α-Mannosidase	pH 4.2, 0.1 M MES, 1.0 M NaCl	SDC	2000	Glycine[g] ethyl ester	8/	—	12
Yeast enolase	0.050 M MES, 1 mM MgCl$_2$, 0.01 mM EDTA, pH 6.1	EDC, CMC[u]	2000	—	—	—[v]	13
3-Phosphoglycerate kinase	0.1 M phosphate, pH 6.1, 17°C	CMC	2000	Nitrotyrosine[v] ethyl ester	1	—	14
cAMP-dependent protein kinase	pH 6.5, 0.050 M MES, 23°C	EDC		Glycine ethyl ester	1.7[w]	—[x]	15

			Time	Nucleophile			Ref.
Human Fc fragment	pH 4.75	EDC	—	Glycine ethyl ester	25	—[y]	16
Pancreatic phospholipase A$_2$	0.25 M cacodylate, pH 5.5	EDC	—	Semicarbazide	13/15[z]	—[aa]	17
Spinach plastocyanin	pH 3.5	EDC	—	Semicarbazide	14/15[z]	—[aa]	17
	pH 6.0, 23°C borate	EDC	—	Ethylenediamine	4.3/16	—	18
Mitochondrial F$_1$-ATPase	0.05 M tri-ethanolamine H$_2$SO$_4$, pH 7.0	Dicyclohexyl-carbodiimide [bb]	6	—	2[cc]	—	19
Mitochondrial transhydrogenase	1 mM Tricine, pH 7.0 with 0.1 M choline chloride and 2% MeOH	DCC	—	—	—	—[dd]	20
Lysozyme	pH 5.0	EDC	3.5[ee]	Ethylenediamine, ethanolamine, 4-(5)-(aminomethyl) imidazole, histamine, D-glucosamine, methylamine	1	—	21
Restriction endonuclease Eco R$_1$	0.1 M triethanolamine pH 7.0, 2.0 M KCl 20°C	CMC	—	—[ff]	—	—	22
Thylakoid membrane proteins	pH 7.5 (HEPES)	DDC	—	Glycine ethyl ester	—	—	23
		EDC	—	Glycine ethyl ester	—	—	

a   It is generally necessary to add dilute HCl (0.2 M) during the course of the reaction to maintain the pH at 4.75.

b   BCD.

c   Reaction time not given.

d   Excess may result from autolysis of trypsin preparations which would create "new" free carboxyl groups.

e   EDC.

f   Reactions are generally terminated by dilution into cold sodium acetate (1.0 M, pH 3.5—5.5).

g   Determined by incorporation of [^{14}C] glycine ethyl ester.

h   1-Hr reaction terminated with 1.0 M acetate, pH 4.75.

i   In 7.5 M urea.

Table 2 (continued)
THE MODIFICATION OF CARBOXYL GROUPS IN PROTEINS

j   Note the interesting difference in extent of modification which is dependent upon nucleophile used. There is also an interesting difference in the time course of modification.

k   In the presence and absence of the competitive inhibitor, benzamidine.

l   Tyrosyl residues regenerated in 0.5 M hydroxylamine, pH 7.1 with no effect on the EDC/glycinamide changes in catalytic activity.

m   Complete modification of the carboxyl groups was achieved in 6.0 M guanidine with either L-argininamide or glycine methyl ester. After reduction and carboxymethylation, approximately 20% of the carboxyl groups are unreactive with either nucleophile. There is a further decrease in modification with the reduced and cyanoethylated derivative.

n   Reaction at pH 4.0 results in apparent quantitative modification of carboxyl groups.

o   Tyrosine modification is acid-stable but reversed in 0.5 M hydroxylamine, pH 7.0, 5 hr, 25°C. "Activated" papain irreversibly modified at active site cysteinyl residue (Cys-25) by EDC while mercuripapain is not.

p   Optimal inactivation occurred at pH 5.5—6.0 with marked decrease in extent of inactivation at more alkaline pH.

q   CMC.

r   Isolated peptide containing modified glutamic acid residue using affinity chromatography with antinitrotyrosyl γ-globulin.

s   Inactivation not dependent upon addition of nucleophile but rate is greatly enhanced.

t   Determined by incorporation of [^{14}C] from [Metho-^{14}C] CMC. Also, determined from the extent of incorporation of N-(2,4-dinitrophenyl)-ethylene diamine (spectrophometry); ϵ = 15,000 M^{-1} cm^{-1}.

u   These investigators also used 1-ethyl-(4-azonia-4,4-dimethylpentyl)-carbodiimide (1-ethyl-3-(3-dimethylaminopropyl)-carbodiimide). This reagent was more effective than either EDC or CMC (least active) in the inactivation of the enzyme under these conditions.

v   Reaction at cysteine and tyrosine excluded by amino acid analysis after acid hydrolysis.

w   Determined γ-glu-gly after proteolysis with trypsin (2X), pronase, carboxypeptidases A and B, and leucine aminopeptidase.

x   Direct determination not performed. Reaction at pH 8.0 (carbodiimide is more specific for phenolic hydroxyl at alkaline pH) did not result in loss of catalytic activity.

y   Tyrosine modification excluded by amino acid analysis after acid hydrolysis.

z   Radiolabeled semicarbazide (synthesized from ^{14}C cyanate) incorporation.

aa Tyrosine modification did occur. Tyrosine regenerated with neutral hydroxylamine. These investigators stressed the need to keep hydroxylamine exposure as brief as possible to avoid side reactions such as peptide bond cleavage or deamidation.

bb DCC.

cc From incorporation of [^{14}C] carbodiimide.

dd Interchain cross-linking of transhydrogenase dimer occurred under these conditions.

ee Obtained specific modification of Asp-101 by using low molar excess of carbodiimide. Extent of modification somewhat independent of amine (nucleophile) used. These investigators speculate that increased specificity is a reflection of binding of carbodiimide to substrate binding site close to Asp-101. These investigators did purify reaction products to obtain selectively modified protein derivatives.

ff Rate and extent of inactivation not changed by the addition of glycine ethyl ester.

References for Table 2

1. **Hoare, D. G. and Koshland, D. E., Jr.,** A procedure for the selective modification of carboxyl groups in proteins, *J. Am. Chem. Soc.,* 88, 2057, 1966.
2. **Radhakrishnan, T. M., Walsh, K. A., and Neurath, H.,** Relief by modification of carboxylate groups of the calcium requirement for the activation of trypsinogen, *J. Am. Chem. Soc.,* 89, 3059, 1967.
3. **Abita, J. P., Maroux, S., Delaage, M., and Lazdunski, M.,** The reactivity of carboxyl groups in chymotrypsinogen, *FEBS Lett.,* 4, 203, 1969.
4. **Carraway, K. L., Spoerl, P., and Koshland, D. E., Jr.,** Carboxyl group modification in chymotrypsin and chymotrypsinogen, *J. Mol. Biol.,* 42, 133, 1969.
5. **Lin, T.-Y. and Koshland, D. E., Jr.,** Carboxyl group modification and the activity of lysozyme, *J. Biol. Chem.,* 244, 505, 1969.
6. **Eyl, A. W., Jr. and Inagami, T.,** Identification of essential carboxyl groups in the specific binding site of bovine trypsin by chemical modification, *J. Biol. Chem.,* 246, 738, 1971.
7. **Frater, R.,** Reactivity of carboxyl groups in modified proteins, *FEBS Lett.,* 12, 186, 1971.
8. **Johnson, P. E., Stewart, J. A., and Allen, K. G. D.,** Specificity of α-chymotrypsin with exposed carboxyl groups blocked, *J. Biol. Chem.,* 251, 2353, 1976.
9. **Perfetti, R. B., Anderson, C. D., and Hall, P. L.,** The chemical modification of papain with 1-ethyl-3(3-dimethylaminopropyl) carbodiimide, *Biochemistry,* 15, 1735, 1976.
10. **Pho, D. B., Roustan, C., Tot, A. N. T., and Pradel, L.-A.,** Evidence for an essential glutamyl residue in yeast hexokinase, *Biochemistry,* 16, 4533, 1977.
11. **Ariki, M. and Fukui, T.,** Modification of rabbit muscle phosphorylase b by a water-soluble carbodiimide, *J. Biochem. (Tokyo),* 83, 183, 1978.
12. **Paus, E.,** Reaction of α-mannosidase from *Phaseolus vulgaris* with group-specific reagents. Essential carboxyl groups, *Biochim. Biophys. Acta,* 526, 507, 1978.
13. **George, A. L., Jr. and Borders, C. L., Jr.,** Essential carboxyl residues in yeast enolase, *Biochem. Biophys. Res. Commun.,* 87, 59, 1979.
14. **Desvages, G., Roustan, C., Fattoum, A., and Pradel, L.-A.,** Structural studies on yeast 3-phosphoglycerate kinase. Identification by immunoaffinity chromatography of one glutamyl residue essential for 3-phosphoglycerate kinase activity. Its location in the primary structure, *Eur. J. Biochem.,* 105, 259, 1980.
15. **Matsuo, M., Huang, C.-H., and Huang, L. C.,** Modification and identification of glutamate residues at the arginine recognition site in the catalytic subunit of adenosine 3′:5′-cyclic monophosphate-dependent protein kinase of rabbit skeletal muscle, *Biochem. J.,* 187, 371, 1980.
16. **Vivanco-Martinez, F., Bragado, R., Albar, J. P., Juarez, C., and Ortiz-Masllorens, F.,** Chemical modification of carboxyl groups in human Fcγ fragment: structural role and effect on the complement fixation, *Mol. Immunol.,* 17, 327, 1980.
17. **Fleer, E. A. M., Verheij, H. M., and de Haas, G. H.,** Modification of carboxylate groups in bovine pancreatic phospholipase A₂. Identification of aspartate-49 as Ca²⁺-binding ligand, *Eur. J. Biochem.,* 113, 283, 1981.
18. **Burkey, K. O. and Gross, E. L.,** Effect of carboxyl group modification on redox properties and electron donation capability of spinach plastocyanin, *Biochemistry,* 20, 5495, 1981.
19. **Pennington, R. M. and Fisher, R. R.,** Dicyclohexylcarbodiimide modification of bovine heart mitochondrial transhydrogenase, *J. Biol. Chem.,* 256, 8963, 1981.
20. **Esch, F. S., Böhlen, P., Otsuka, A. S., Yoshida, M., and Allison, W. S.,** Inactivation of the bovine mitochondrial F₁-ATPase with dicyclohexyl [¹⁴C] carbodiimide leads to the modification of a specific glutamic acid residue in the β-subunit, *J. Biol. Chem.,* 256, 9084, 1981.
21. **Yamada, H., Imoto, T., Fujita, K., Okazaki, K., and Motomura, M.,** Selective modification of aspartic acid-101 in lysozyme by carbodiimide reaction, *Biochemistry,* 20, 4836, 1981.
22. **Woodhead, J. L. and Malcolms, D. B.,** The essential carboxyl group in restriction endonuclease Eco RI, *Eur. J. Biochem.,* 120, 125, 1981.
23. **Laszlo, J. A., Millner, P. A., and Dilley, R. A.,** Light-dependent chemical modification of thylakoid membrane proteins with carboxyl-directed reagents, *Arch. Biochem. Biophys.,* 215, 571, 1982.

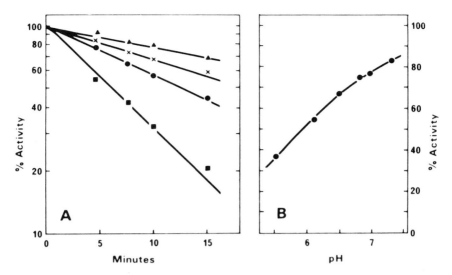

FIGURE 5.   Inactivation of hexokinase by 1-cyclohexyl-3-[2-morpholinyl-(4)-ethyl]carbodiimide metho-*p*-toluenesulfonate(CMC). Panel A shows the effect of CMC concentration with 36 μ*M* enzyme in 0.01 *M* phosphate buffer, pH 6.0, at 20°C. The CMC concentration was (▲) 0.015 *M*, (x) 0.03 *M*, (●) 0.05 *M*, and (■) 0.10 *M*. Panel B shows the effect of pH(phosphate buffers) on the reaction with 0.05 *M* CMC(10 min reaction time). (From Pho, D. B., Roustan, C., Tot, A.N.T., and Pradel, L.–A., *Biochemistry,* 16, 4533, 1977. With permission.)

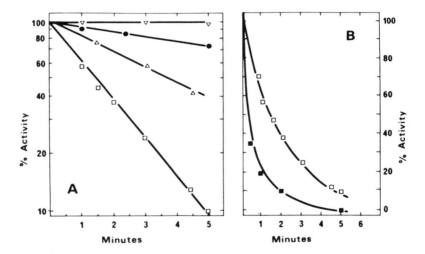

FIGURE 6.   The effect of 1-cyclohexyl-3-[2-morpholinyl-(4)-ethyl]carbodiimide metho-*p*-toluenesulfonate(CMC) and nitrotyrosine ethyl ester(NTEE) on hexokinase activity. The concentration of enzyme was 36 μ*M* in 0.01 *M* phosphate buffer, pH 6.0. Panel A: (▽), 0.03 *M* NTEE alone; (●), 0.05 *M* CMC alone; (□), 0.05 *M* CMC + 0.03 *M* NTEE; and (△), 0.025 *M* glucose + 0.005 *M* ADP-Mg + 0.05 *M* CMC + 0.03 *M* NTEE. Panel B: with different concentrations of CMC; (□) 0.05 *M* CMC + 0.03 *M* NTEE and (■) 0.1 *M* CMC + 0.03 *M* NTEE. (From Pho, D. B., Roustan, C., Tot, A.N.T., and Pradel, L.–A., *Biochemistry,* 16, 4533, 1977. With permission.)

matography as shown in Figure 10 and Figure 11. The modification of pancreatic phospholipase A$_2$ (Figure 12) provides a particularly useful example of the effect of pH on carbodiimide-mediated modification. Note that the inactivation is much more rapid at pH 3.5 than at pH 5.5.

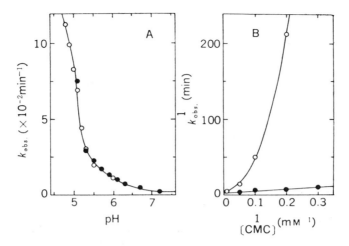

FIGURE 7. The inactivation of phosphorylase with 1-cyclohexyl-3-[2-morpholinyl-(4)-ethyl]carbodiimide metho-*p*-toluenesulfonate(CMC). Panel A shows the pH dependence of the observed pseudo first-order rate constant (k_{obs}) for the inactivation of phosphorylase by 20 mM CMC. The plot was obtained from two separate experiments, each represented by ○ and ●. Panel B shows a double reciprocal plot of k_{obs} vs. the concentration of CMC in the presence (●) and absence (○) of 50 mM glycine ethyl ester. (From Ariki, M. and Fukui, T., *J. Biochem.*, 83, 183, 1978. With permission.)

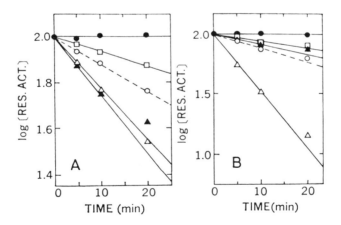

FIGURE 8. The inactivation of phosphorylase with 1-cyclohexyl-3-[2-morpholinyl-(4)-ethyl]carbodiimide metho-*p*-toluenesulfonte(CMC) in the presence of various nucleophiles. Panel A shows the effect of various amines on the inactivation rate of phosphorylase b with 10 mM CMC. (●), No reagent; (○), CMC alone; (△), CMC + 50 nM glycine ethyl ester; (□), CMC + 50 mM 2-amino-2-deoxyglucose; and (▲), CMC + 50 mM glucosyl amine. Panel B shows the effect of amino analogs of glucose on the inactivation rate of phosphorylase b with 20 mM CMC. (●), No reagent added; (○), CMC alone; (△), CMC + 50 mM glycine ethyl ester; (▲) CMC + 50 mM 3-amino-3-deoxyglucose; and (□), CMC + 50 mM 6-amino-6-deoxyglucose. (From Ariki, M. and Fukui, T., *J. Biochem.*, 83, 183, 1978. With permission.)

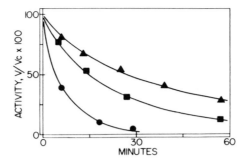

FIGURE 9. The inactivation of yeast enolase by various carbodiimides. Enolase(1 μ*M*) was modified by 20 m*M* carbodiimide in 50 m*M* MES, 1 m*M* MgCl$_2$, 0.01 m*M* EDTA, pH 6.1, at 25°C. The carbodiimides used were (●) 1-ethyl-3-(4-azonia-4,4-dimethylpentyl)-carbodiimide iodide, (■) 1-ethyl-3-(3-dimethylaminopropyl)-carbodiimide, and (▲) 1-cyclohexyl-2-[3-morpholinyl-(4)-ethyl]-carbodiimide metho-*p*-toluenesulfonate. (From George, A. L., Jr. and Borders, C. L., Jr., *Biochem. Biophys. Res. Commun.*, 87, 59, 1979. With permission.)

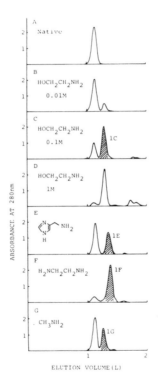

FIGURE 10. Ion-exchange chromatography of the product obtained from the reaction of egg-white lysozyme with various amines in the presence of 1-ethyl-3-[3-(dimethylamino)propyl]-carbodiimide hydrochloride. The modifications were performed at ambient temperature at pH 5.0 (maintained by the addition of HCl). The modified proteins were dialyzed exhaustively against water and then analyzed by ion-exchange chromatography on a 1 × 65 cm Bio-Rex 70 column using a linear gradient from 0.02 *M* borate, pH 10.0 to 0.02 *M* borate-0.15 *M* NaCl, pH 10.0. A, native lysozyme; B, reaction with 0.01 *M* ethanolamine; C, reaction with 0.1 *M* ethanolamine; D, reaction with 1 *M* ethanolamine; E, reaction with 4(5)-(aminomethyl)imidazole; and G, reaction with methylamine. (From Yamada, H., et al., *Biochemistry*, 20, 4836, 1981. With permission.)

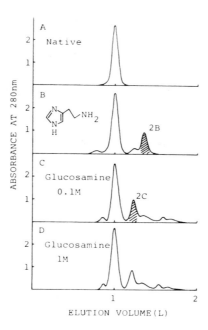

FIGURE 11. Ion-exchange chromatography of the product of the reaction of egg-white lysozyme with various amines in the presence of 1-ethyl-3-[3-(dimethylamino)propyl]-carbodiimide hydrocholoride. The reactions were performed as described under Figure 10 and subjected to chromatographic analysis on a 1 × 65 cm column of Bio-Rex 70 using a linear gradient from 0.1 M phosphate, pH 7.0, to 0.4 M phosphate, pH 7.0. A, native lysozyme; B, reaction with histamine; C, reaction with 0.1 M D-glucosamine; and D, reaction with 1 M D-glucosamine.

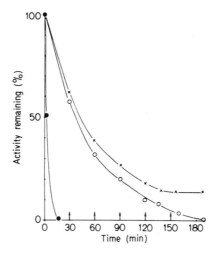

FIGURE 12. Modification of carboxylate groups in bovine pancreatic phospholipase A$_2$. Shown is the loss of enzyme activity vs reaction time. Protein (5 mg/mℓ) was dissolved in 1 M semicarbazide. At the indicated time intervals (arrows), 5 mg 1-ethyl-3-(N,N-dimethyl)amino-propyl carbodiimide was added. (●), reaction at pH 3.5 (semicarbizide-HCl buffer) and (○), reaction at pH 5.5 (0.25 M cacodylate buffer); both reactions without Ca^{2+}. (x) Reaction at pH 5.5 in the presence of Ca^{2+} ions. The reaction was started by the addition of carbodiimide to 0.1 M. (From Fleer, E.A.M., Verheij, H. M., and de Haas, *Eur. J. Biochem.*, 113, 283, 1981. With permission.)

REFERENCES

1. **Herriott, R. M.,** Reactions of native proteins with chemical reagents, *Adv. Protein Chem.,* 3, 169, 1947.
2. **Hoare, D. G. and Koshland, D. E., Jr.,** A procedure for the selective modification of carboxyl groups in proteins, *J. Am. Chem. Soc.,* 88, 2057, 1966.
3. **Takahaski, K., Stein, W. H., and Moore, S.,** The identification of a glutamic acid residue as part of the active site of ribonuclease T_1, *J. Biol. Chem.,* 242, 4682, 1967.
4. **Erlanger, B. F., Vratsanos, S. M., Wassermann, M., and Cooper, A. G.,** Specific and reversible inactivation of pepsin, *J. Biol. Chem.,* 240, PC3447, 1965.
5. **Gross, E. and Morell, J. L.,** Evidence for an active carboxyl group in pepsin, *J. Biol. Chem.,* 241, 3638, 1966.
6. **Tang, J.,** Specific and irreversible inactivation of pepsin by substrate-like epoxides, *J. Biol. Chem.,* 246, 4510, 1971.
7. **Chen, K. C. S. and Tang, J.,** Amino acid sequence around the epoxide-reactive residues in pepsin, *J. Biol. Chem.,* 247, 2566, 1972.
8. **Eshdat, Y., McKelvy, J. F., and Sharon, N.,** Identification of aspartic acid 52 as the point of attachment of an affinity label in hen egg white lysozyme, *J. Biol. Chem.,* 248, 5892, 1973.
9. **Eshdat, Y., Dunn, A., and Sharon, N.,** Chemical conversion of aspartic acid 52, a catalytic residue in hen egg white lysozyme, to homoserine, *Proc. Natl. Acad. Sci. U.S.A.,* 71, 1658, 1974.
10. **Wilcox, P. E.,** Esterification, *Meth. Enzymol.,* 11, 605, 1967.
11. **Wilcox, P. E.,** Esterification, *Meth. Enzymol.,* 25, 596, 1972.
12. **Rajagopalan, T. G., Stein, W. H., and Moore, S.,** The inactivation of pepsin by diazoacetylnorleucine methyl ester, *J. Biol. Chem.,* 241, 4295, 1966.
13. **Lundblad, R. L. and Stein, W. H.,** On the reaction of diazoacetyl compounds with pepsin, *J. Biol. Chem.,* 244, 154, 1969.
14. **Stein, W. H.,** Chemical studies on purified pepsin, in *Structure-Function Relationships of Proteolytic Enzymes,* Desnuelle, P., Neurath, H., and Ottesen, M., Eds., Munksgard, Copenhagen, 1970, 253.
15. **Leary, R. and Shaw, E.,** Inactivation of cathepsin B_1 by diazomethyl ketones, *Biochem. Biophys. Res. Commun.,* 79, 926, 1977.
16. **Widmer, F. and Viswanatha, T.,** Possible side-reactions with diazocarbonyl dipeptide esters as protein modifying reagents, *Carlsberg Res. Commun.,* 45, 149, 1980.
17. **Parsons, S. M., Jao, L., Dahlquist, F. W., Borders, C. L., Jr., Groff, T., Racs, J., and Raftery, M. A.,** The nature of amino acid side chains which are critical for the activity of lysozyme, *Biochemistry,* 8, 700, 1969.
18. **Parsons, S. M. and Raftery, M. A.,** The identification of aspartic acid residue 52 as being critical to lysozyme activity, *Biochemistry,* 8, 4199, 1969.
19. **Paterson, A. K. and Knowles, J. R.,** The number of catalytically essential carboxyl groups in pepsin. Modification of the enzyme by trimethyloxonium fluoroborate, *Eur. J. Biochem.,* 31, 510, 1972.
20. **Gagnon, C., Viveros, O. H., Diliberto, E. J., Jr., and Axelrod, J.,** Enzymatic methylation of carboxyl groups of chromaffin granule membrane proteins, *J. Biol. Chem.,* 253, 3778, 1978.
21. **Woodward, R. B., Olofson, R. A., and Mayer, H.,** A new synthesis of peptides, *J. Am. Chem. Soc.,* 83, 1010, 1961.
22. **Woodward, R. B. and Olofson, R. A.,** The reaction of isoxazolium salts with nucleophiles, *Tetrahedron,* Suppl. 7, 415, 1966.
23. **Bodlaender, P., Feinstein, G., and Shaw, E.,** The use of isoxazolium salts for carboxyl group modification in proteins. Trypsin, *Biochemistry,* 8, 4941, 1969.
24. **Saini, M. S. and Van Etten, R. L.,** An essential carboxylic acid group in human prostate acid phosphatase, *Biochim. Biophys. Acta,* 568, 370, 1979.
24. **Arana, J. L. and Vallejos, R. H.,** Two different types of essential carboxyl groups in chloroplast coupling factor, *FEBS Lett.,* 123, 103, 1981.
26. **Dinur, D., Kantrowitz, E. R., and Hajdu, J.,** Reaction of Woodward's Reagent K with pancreatic porcine phospholipase A_2: modification of an essential carboxylate residue, *Biochem. Biophys. Res. Commun.,* 100, 785, 1981.
27. **Banks, T. E., Blossey, B. K., and Shafer, J. A.,** Inactivation of α-chymotrypsin by a water-soluble carbodiimide, *J. Biol. Chem.,* 244, 6323, 1969.
28. **Perfetti, R. B., Anderson, C. D., and Hall, P. L.,** The chemical modification of papain with 1-ethyl-3(3-dimethylaminopropyl) carbodiimide, *Biochemistry,* 15, 1735, 1976.
29. **Carraway, K. L. and Koshland, D. E., Jr.,** Reaction of tyrosine residues in proteins with carbodiimide reagents, *Biochim. Biophys. Acta,* 160, 272, 1968.
30. **Carraway, K. L. and Koshland, D. E., Jr.,** Carbodiimide modification of proteins, *Meth. Enzymol.,* 25, 616, 1972.

31. **Lewis, S. D. and Shafer, J. A.,** Conversion of exposed aspartyl and glutamyl residues in proteins to asparaginyl and glutaminyl residues, *Biochim. Biophys. Acta,* 303, 284, 1973.
32. **Wang, T.-T. and Young, N. M.,** Modification of aspartic acid residues to induce trypsin cleavage, *Analyt. Biochem.,* 91, 696, 1978.
33. **Pho, D. B., Roustan, C., Tot, A. N. T., and Pradel, L.-A.,** Evidence for an essential glutamyl residue in yeast hexokinase, *Biochemistry,* 16, 4533, 1977.
34. **Ariki, M. and Fukui, T.,** Modification of rabbit muscle phosphorylase b by a water-soluble carbodiimide, *J. Biochem.,* 83, 183, 1978.
35. **George, A. L., Jr. and Borders, C. L., Jr.,** Essential carboxyl residues in yeast enolase, *Biochem. Biophys. Res. Commun.,* 87, 59, 1979.
36. **Yamada, H., Imoto, T., Fujita, K., Okazaki, K., and Motomura, M.,** Selective modification of aspartic acid-101 in lysozyme by carbodiimide reaction, *Biochemistry,* 20, 4836, 1981.
37. **Fleer, E. A. M., Verheij, H. M., and de Haas,** Modification of carboxylate groups in bovine pancreatic phospholipase A_2. Identification of aspartate-49 as Ca^{2+}-binding ligand, *Eur. J. Biochem.,* 113, 283, 1981.

Chapter 5

THE CHEMICAL CROSS-LINKING OF PEPTIDE CHAINS

The formation of either intramolecular or intermolecular covalent cross-links between amino acid residues in proteins is proving to be an extremely valuable tool in biochemistry with particular use in the study of protein-protein interactions. Naturally occurring inter- and intramolecular cross-links are commonly found in proteins, the most common being the disulfide bond. Other examples exist including the transglutaminase-catalyzed formation of a peptide bond between the γ-carboxyl groups of glutamic acid and the ε-amino groups of lysine.[1,2] There are also the extremely complex cross-links found in collagen.[3,4]

The formation of and subsequent determination of the sites of intrachain or interchain covalent cross-links provide a direct approach to the study of the folding and interaction of polypeptide chains in solution. A number of important questions can be addressed through the introduction of covalent cross-links into a protein. The formation of intramolecular cross-links within a single polypeptide chain can provide information regarding inter-residue distances and the relationships between various domains in a protein. Care must be used in the interpretation of such results with respect to the effect of such cross-linking on catalytic function or biological function. In order to draw meaningful conclusions one must know the effect of the reaction of amino acid residues with monofunctional analogues of the bifunctional reagent. In other words, derivatives formed by the reagent X-R-R-X or X-R-R-Y should be compared with the reaction products of RY and/or RX with the same residue(s).

One of the more interesting examples of intramolecular cross-linking has involved the use of 1,3-dibromoacetone (Figure 1). Either methylene carbon would be subject to nucleophilic attack. Husain and Lowe[5] obtained intramolecular cross-linking between the active site cysteine and the active site histidine in ficin and stem bromelain. These results suggested that the histidine residue is with 5Å of the active site cysteine. No intermolecular cross-linking occurred under the reaction conditions used by these investigators (0.05 *M* sodium acetate, 0.0001 *M* EDTA, pH 5.6; the dibromoacetone was dissolved in acetone). Spatial proximity is of critical importance since treatment of glyceraldehyde-3-phosphate dehydrogenase with 1,3-dibromoacetone resulted in the modification of cysteine but not histidine.[6] These investigators reported the synthesis of the radiolabeled derivative.

The synthesis of bifunctional inhibitors for pepsin has also been reported[7] (Figure 2).

1,1-Bis(diazoacetyl)-2-phenylethane was found to be a potent inhibitor of pepsin (0.008 *M* sodium acetate, 0.0002 *M* CuSO$_4$, pH 5.0). Little, if any, intermolecular cross-linking occurred under these conditions, and the data presented were consistent with the cross-linkage of two residues at the active site. The reaction probably involves two carboxyl groups[8] (see Chapter 4).

An interesting pair of reagents (Figure 3) which cross-link proteins by consecutive Michael reactions have been described.[9] These investigators made the point that since the cross-linkage reaction is driven by consecutive Michael additions, eventually the most thermodynamically stable cross-link will be established, which can be subsequently stabilized by reduction of the nitro function with sodium dithionite. These investigators explored the reaction of pancreatic ribonuclease with 2-(*p*-nitrophenyl) allyl-4-nitro-3-carboxyphenyl sulfide (twofold molar excess with respect to ribonuclease in 0.1 *M* sodium phosphate, pH 10.5 at 37°C, 36 hr, cross-link stabilized by sodium dithionite fivefold molar excess). No reaction occurred at pH 8.0. Analysis of the reaction mixture showed 61% monomer, 21% dimer, 10% trimer, and a trace of tetramer. The monomer fraction was characterized; the predominant cross-links occurred between lysine-7 and lysine-37 and between lysine-31 and lysine-41.

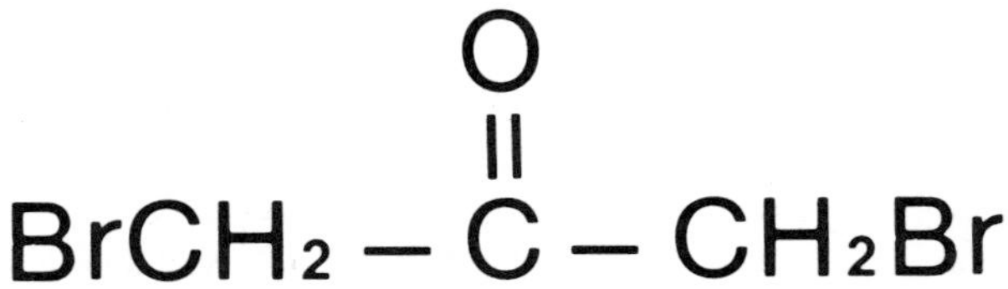

$$\underset{\text{BrCH}_2 - \overset{\displaystyle \overset{O}{\parallel}}{\text{C}} - \text{CH}_2\text{Br}}{}$$

FIGURE 1.   The structure of 1,3-dibromoacetone.

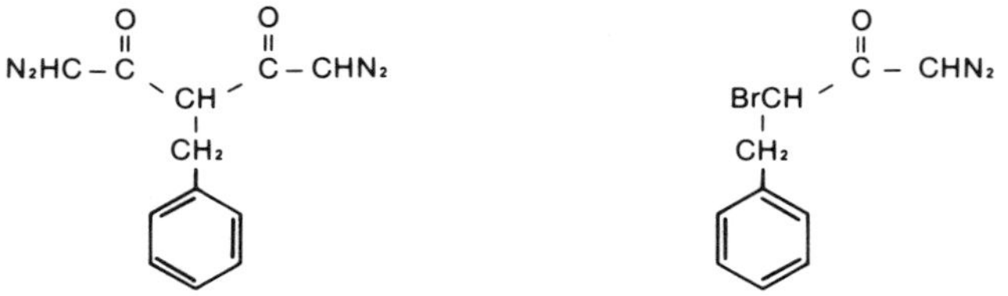

FIGURE 2.   The structures of two bifunctional inhibitors of pepsin. On the left is shown the structure for 1,1-bis(diazoacetyl)-2-phenylethane and on the right is the structure for 1-diazoacetyl-1-bromo-2-phenylethane.

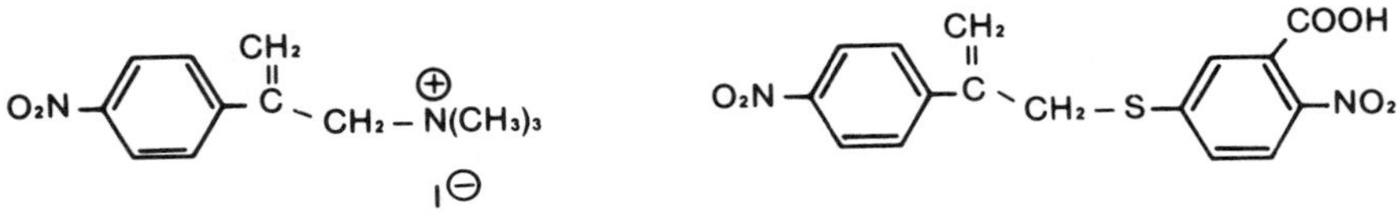

FIGURE 3.   The structure of 2-(*p*-nitrophenyl)allyltrimethylammonium iodide (on the left) and the structure of 2-(*p*-nitrophenyl)-allyl-4-nitro-3-carboxyphenyl sulfide (on the right).

Care must be taken in the size analysis of intramolecularly linked species in any denaturing medium (i.e., sodium dodecylsulfate, guanidine hydrochloride, etc.) since, unless the cross-linking reagent is cleaved by reduction, the intramolecularly cross-linked protein will not denature properly and will probably give falsely low molecular weight results.[10,11]

Intramolecular cross-linking can be enhanced by the following reaction conditions:

1.   Low protein concentration (<0.1 mg/mℓ)
2.   High net charge on protein
3.   High ratio of protein reactive sites to reagent concentration

The remainder of our consideration will involve intermolecular cross-linking, which includes the cross-linking of identical protomers to form homopolymers (e.g., cross-linkage of identical subunits in an oligomeric protein).[12,13] Cross-linkage to form heteropolymers includes studies on protein-protein interactions (this could result in homopolymers in self-associating systems),[14,15] studies on multienzyme complexes,[16-18] and protein-ligand interactions with cell membrane receptors.[19-24]

The following discussion will focus on the various reagents which have been used for intermolecular protein cross-linking studies.

Glutaraldehyde (Figure 4) should cross-link proteins with the formation of α,ω-Schiff bases, which should be a readily reversible process in the absence of reduction of the Schiff

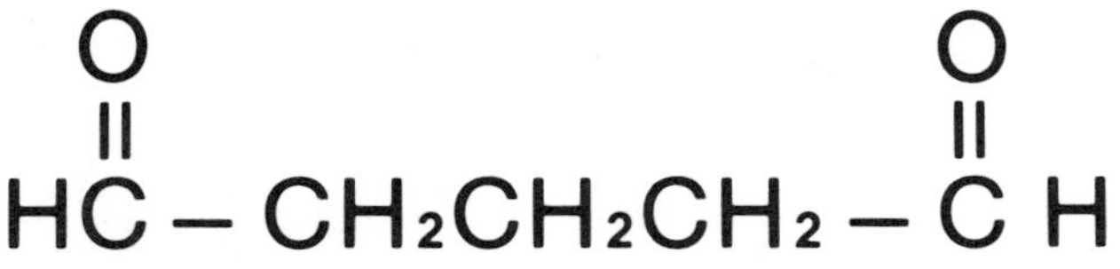

$$\underset{\displaystyle HC - CH_2CH_2CH_2 - CH}{\overset{\displaystyle O \qquad\qquad\qquad\qquad O}{\underset{\displaystyle \parallel \qquad\qquad\qquad\qquad \parallel}{}}}$$

FIGURE 4.   The structure of glutaraldehyde.

bases with a reducing agent such as sodium borohydride or sodium cyanoborohydride. This is not the case, as summarized by Richards and Knowles.[25] These investigators noted that the reaction of proteins with glutaraldehyde was essentially irreversible even without reduction and that, in the absence of reduction, there was loss of lysine on amino acid analysis following acid hydrolysis. These investigators proposed the formation of a complex reagent resulting from aldol condensation of glutaraldehyde which would then react with the protein. More recent studies have presented an alternative structure for glutaraldehyde.[26] It seems clear from these studies that the chemistry of the reaction of glutaraldehyde is complex. In recent years, the use of glutaraldehyde received considerable attention in the study of the properties of protein crystals in solution.[27-31] The rationale in these studies has been to show that the properties of a protein in crystalline form are similar to those for the protein in solution.[31] Although glutaraldehyde has a fair degree of specificity for the ϵ-amino group of lysine, reaction has also occurred with other nucleophilic functional groups in proteins such as the sulfhydryl group of cysteine, the imidazole ring of histidine, and the phenolic hydroxyl group of tyrosine.[32]

The reaction of glutaraldehyde with proteins proceeds quite rapidly at alkaline pH as demonstrated by early studies[33] on reaction with chymotrypsin (Figure 5).

More recent studies[34] have utilized glutaraldehyde to measure the rate of reconstitution of porcine muscle lactic dehydrogenase. This enzyme is composed of four subunits (Figure 6) and this study demonstrated that the formation of tetramers from a denatured enzyme preparation parallels the recovery of enzymatic activity (Figure 7).

The next general class of reagents that we will discuss is the homobifunctional imidoesters (Figure 8). This class of reagents was introduced by Singer and co-workers.[35] These reagents have the advantage that the reaction with the protein results in charge preservation of the lysine residue modified. This class of reagents is highly specific for primary amines in the following reactions as shown in Figures 9 and 10. Aspects of the chemistry of this reaction are further discussed in Volume I, Chapter 10. Buffer effects on the reaction have not been extensively investigated except to specify that the use of potential competing nucleophiles (e.g., Tris, imidazole) should be avoided. Most studies have used 0.02 to 0.1 M triethanolamine in the range of pH 8.0 to 9.0. It has been suggested that the amidation reaction is enhanced by the presence of triethanolamine in studies on the reaction of methyl-4-mercaptobutyrimidate.[22] The greatest use of these reagents has been in the study of protomer organization of oligomeric proteins and self-association systems.

The early studies of Davies and Stark with dimethyl suberimidate[12] showed that analysis of the reaction products reveals a set of species with molecular weights equal to integral multiples of the protomer molecular weight. For oligomers composed of identical protomers, the number of products is equivalent to the number of protomers in the oligomer. Figure 11 shows the results obtained with glutaraldehyde-3-phosphate dehydrogenase (four identical subunits), aldolase (four identical subunits), and tryptophan synthetase B protein (two subunits). The results obtained with the catalytic subunit of aspartate transcarbamylase (three subunits, 33,000 daltons) are shown in Figure 12 while the molecular weight plot (log molecular weight vs. mobility) is presented in Figure 13. These investigators also noted that

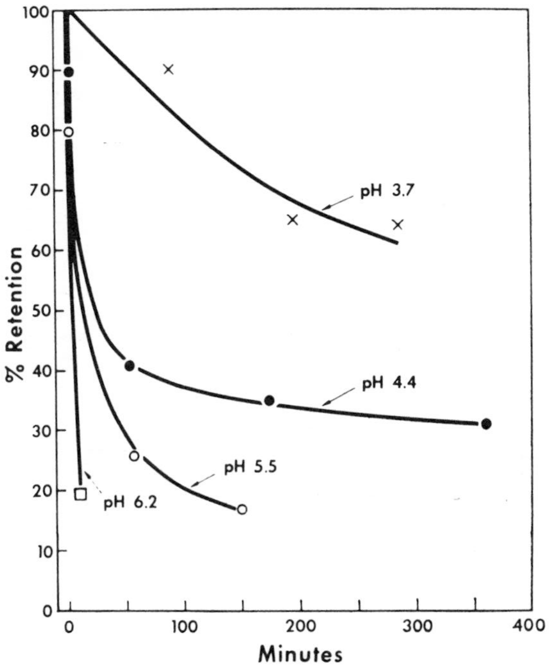

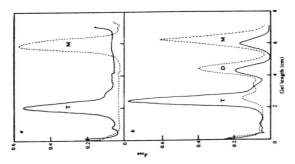

FIGURE 6.   Polyacrylamide gel electrohporesis of cross-linked porcine muscle lactic de-hydrogenase (LDH) in the presence of sodium dodecyl sulfate (SDS). Cross-linking by a 2 min incubation in 0.2 M phosphate, 5 mM EDTA with 0.3 to 0.4 mM glutaraldehyde at 20°C was stopped by addition of 15 to 25 µM SDS. Unreacted bifunctional reagent was inactivated by 3 to 4 mM hydrazine. The solutions were heated for 10 min (100°C) and subsequently stored for 2 hr at 20°C. The protein solutions were concentrated by 24-hr dialysis vs. 20% polyethylene glycol prior to electrophoresis on 5% polyacrylamide gels in the presence of SDS. The gels were stained with Coomassie blue R 250 and then scanned at 560 nm. Correlation of the bands to the tetrameric(T), dimeric(D), and monomeric(M) fractions of LDH was achieved by calibration with proteins of defined molecular weight. (a), cross-linking of native tetramers (solid line) and SDS-denatured monomers (dashed line); monomer concentration 81 nM. (b), Cross-linking of reactivating LDH after acid dissociation at pH 2.3. Reactivation at 20°C by dilution at pH 7.6. The final concentration of monomer was 343 nM. The dashed line indicates cross-linking after 210 sec while the solid line indicates 1 hr of reactivation and reassociation. (From Hermann, R., Rudolph, R., and Jaenicke, R., *Nature (London)*, 277, 243, 1979. With permission.)

FIGURE 5.   The effect of pH on the inactivation of α-chy-motrypsin by glutaraldehyde. Reaction mixtures were 2.3% in glutaraldehyde and 0.2% in α-chymotrypsin at 0°C. For pH values of 3.7, 4.4, and 5.5, 0.14 N acetate was employed as buffer. The buffer for pH 6.2 was 0.034 M phosphate. (From Jansen, E. F., Tomimatsu, Y., and Olson, A. C., *Arch. Biochem. Biophys.*, 144, 394, 1971. With permission.)

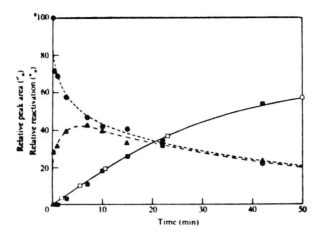

FIGURE 7. Determination of the kinetics of reassociation of porcine muscle lactic dehydrogenase by cross-linking. The experimental conditions are described in the caption to Figure 6. Reassociation (20°C) at 170 nM monomer concentration was analyzed by withdrawing portions from the reactivation mixture at defined times and fixation of the association products by cross-linking with glutaraldehyde. Percentage of monomers (●), dimers (▲), and tetramers (■) is defined by the respective peak areas relative to their total amount. Reactivation (□), as determined by enzymatic assay, was calculated relative to the final extent of reactivation (75% after 6 days). (From Hermann, R., Rudolph, R., and Jaenicke, R., *Nature (London)*, 277, 243, 1979. With permission.)

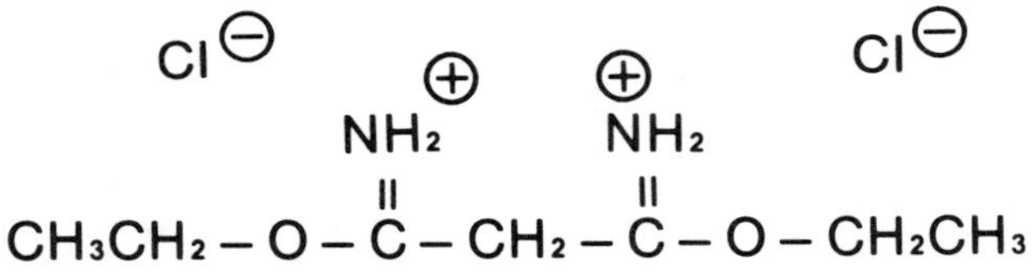

$$Cl^{\ominus} \qquad \overset{\oplus}{NH_2} \quad \overset{\oplus}{NH_2} \qquad Cl^{\ominus}$$
$$CH_3CH_2 - O - \overset{\parallel}{C} - CH_2 - \overset{\parallel}{C} - O - CH_2CH_3$$

FIGURE 8. The structure of a homobifunctional imidoester.

$$2P - NH_2 + CH_3CH_2 - O - \overset{\overset{\oplus}{NH_2}}{\underset{\parallel}{C}} - CH_2 - \overset{\overset{\oplus}{NH_2}}{\underset{\parallel}{C}} - O - CH_2CH_3 \longrightarrow P - NH - \overset{\overset{\oplus}{NH_2}}{\underset{\parallel}{C}} - CH_2 - \overset{\overset{\oplus}{NH_2}}{\underset{\parallel}{C}} - NH - P$$

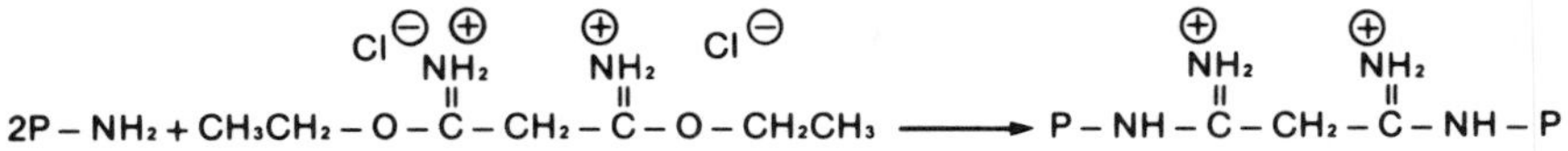

FIGURE 9. The reaction of a homobifunctional imidoester with the primary amino groups in a protein resulting in covalent cross-linking.

$$P - NH_2 + CH_3CH_2 - O - \overset{\overset{\oplus}{NH_2}}{\underset{\parallel}{C}} - CH_2 - \overset{\overset{\oplus}{NH_2}}{\underset{\parallel}{C}} - O - CH_2CH_3 \longrightarrow P - NH - \overset{\overset{\oplus}{NH_2}}{\underset{\parallel}{C}} - CH_2 - \overset{\overset{\oplus}{NH_2}}{\underset{\parallel}{C}} - O - CH_2CH_3$$

$$\downarrow$$

$$P - NH - \overset{\overset{\oplus}{NH_2}}{\underset{\parallel}{C}} - CH_2 - \overset{O}{\underset{\parallel}{C}} - O - CH_2CH_3$$

$$+$$

$$\overset{\oplus}{NH_4} + H_3\overset{\oplus}{O} + CH_3CH_2OH$$

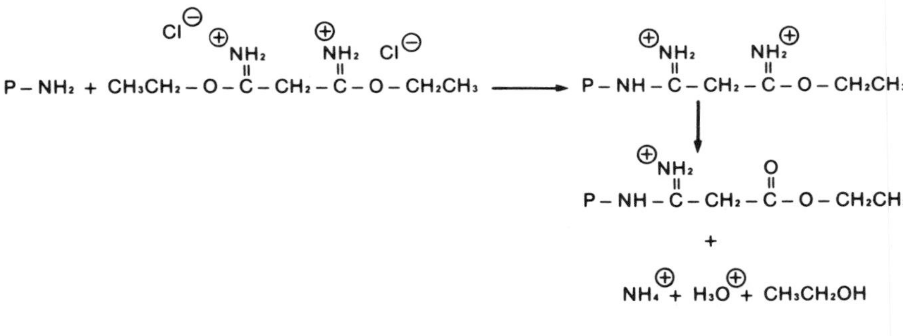

FIGURE 10. The reaction of a homobifunctional imidoester with the primary amino groups in a protein not resulting in covalent cross-linking.

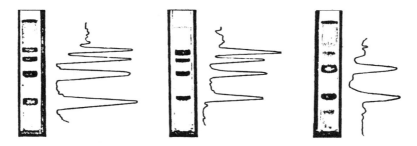

FIGURE 11. Cross-linking studies with glyceraldehyde-3-phosphate dehydrogenase, aldolase and tryptophan synthetase using dimethyl suberimidate. Shown are: left, glyceraldehyde-3-phosphate dehydrogenase (30 μg) cross-linked at 3 mg/mℓ protein and 2 mg/mℓ dimethyl suberimidate; center, aldolase (20 μg) cross-linked at 5 mg/mℓ protein and 0.8 mg/mℓ dimethyl suberimidate; and right, tryptophan synthetase protein (30 μg) cross-linked at 2 mg/mℓ protein and 2 mg/mℓ dimethyl suberimidate. The cross-linking experiments were performed in 0.2 *M* triethanolamine, pH 8.5. Polyacrylamide (5%) gel electrophoresis was performed in the presence of sodium dodecyl sulfate(SDS). The proteins were denatured for 2 hr at 37°C in 1% SDS and 1% β-mercaptoethanol. The gels were stained with Coomassie Blue and densitometer tracing was obtained with a Joyce-Loebl double-beam recording microdensitometer. (From Davies, G. E. and Stark, G. R., *Proc. Natl. Acad. Sci. U.S.A.*, 66, 651, 1970. With permission.)

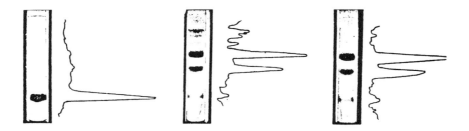

FIGURE 12. Cross-linking studies of aspartate transcarbamylase as a function of protein concentration. The experiments were performed as described in Figure 11. Catalytic subunit of aspartate transcarbamylase: left, untreated (40 μg); center, 40 μg cross-linked at 5 mg/mℓ protein; and right, 40 μg cross-linked at 0.5 mg/mℓ protein. The dimethyl suberimidate concentration was 1 mg/mℓ. (From Davies, G. E. and Stark, G. R., *Proc. Natl. Acad. Sci. U.S.A.*, 66, 651, 1970. With permission.)

reaction at a lower protein concentration can be used to distinguish cross-linking within an oligomer from cross-linking between two or more oligomers (Figure 12). Studies on leucine aminopeptidase from bovine lens[13] also provided a sound basis for the use of bifunctional imidoesters as a probe of the quaternary structure of proteins.

In this study it was emphasized that it is essential to know the effect of reaction with monofunctional reagents on biological activity before considering the effect of cross-linking on biological activity. A recent report in which the effect of methyl acetimidate was compared with dimethylsuberimidate[36] is well worth consideration. In addition, the work of Sinha and Brew[37] is worth further examination. These investigators developed a useful procedure employing the prior trace-labeling of the protein with acetic anhydride. Since reaction with acetic anhydride and imidoester is mutually exclusive, fragmentation and subsequent determination of specific radioactivity at specific lysine residues allows the identification of the site(s) of reaction with bifunctional imidoesters.

Diimidates have been used to study the glycogen phosphorylase system. Reagents of

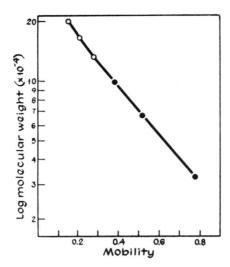

FIGURE 13. Semilogarithmic plot of molecular weight vs. migration relative to bromphenol blue for the covalently linked species produced by the reaction of the catalytic subunit of aspartate transcarbamylase with dimethyl suberimidate. The data were taken from experiments described in Figure 12. (Fom Davies, G. E. and Stark, G. R., *Proc. Natl. Acad. Sci. U.S.A.*, 66, 651, 1970. With permission.)

different length were used to study intermolecular and intramolecular interaction of phosphorylase b.[38] Figure 14 illustrates the effect of AMP on the cross-linkage pattern of phosphorylase b. In panel A, C_d can be viewed as a measure of association between individual dimers of phosphorylase b. Panel B is a plot of reagent (cross-linkage reagent) length vs. r_k which is a measure of total cross-link formation (inter and intra). Figure 15 is from the same study exploring the effect of a number of different ligands on r_k in the presence and absence of AMP ($r_k = k_L/k_o$ where k_L = first-order rate constant for the disappearance of monomer in the presence of a given ligand and k_o = first-order rate constant for the same event in the absence of the ligand — if r_k is less than 1, the ligand inhibits cross-linkage, if r_k is greater than 1, the ligand promotes cross-linkage). Subsequent studies[39] from the same laboratory extended these observations to phosphorylase a and a hybrid phosphorylase ab (Figures 16 and 17), permitting the development of a model (Figure 18) of the various structural states of this highly regulated enzyme.

Disuccinimidyl ester derivatives have also seen considerable use. The introduction of a cross-linking reagent that could be subsequently cleaved, 3,3'-dithiobis(succinimidylpropionate)[40] has proved to be of considerable use in the study of proteins.[21,41-43] This reagent provides for the reversible formation of cross-links according to Figure 19.

The use of bifunctional maleimides is of considerable value in cross-linking between sulfhydryl groups. A particularly useful study in this area is the work of Heilmann and Holzner[44] in which the synthesis and use of a number of bifunctional maleimides for the elucidation of the structure of tryptophan synthetase is reported (Figure 20). One of the more useful derivatives is bis-*N*-maleimido-1,8-octane (Figure 21).

A cleavable bifunctional dimaleimide has been recently reported.[45] The synthesis of the reagent, maleimidomethyl-3-maleimido propionate (Figure 22), is reported as is the use of the reagent to probe spatial relationships in the erythrocyte membrane.

One of the more active developments in this area has been the further use of heterobifunctional reagents for cross-linking of proteins. Of particular interest has been the use of photoactivatable derivatives. An example of this type of derivative is methyl-3-[*p*-azido-

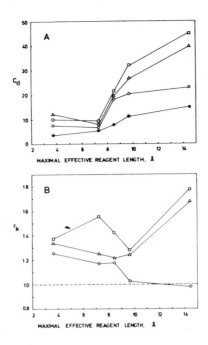

FIGURE 14.   Structural changes in glycogen phosphorylase b as revealed by cross-linking with a homologous series of bifunctional imidates. Shown is the effect of AMP on the cross-link parameters (C_d, the percentage amount of protein found in the trimer and tetramer bands, C_d = (trimer + tetramer)/total × 100; r_k, the rate constant ratio of cross-linking, $r_k = k_L/k_O$ where k_L and k_O are the apparent first-order rate constants of the disappearance of the monomeric band (as detected by gel electrophoresis) in the presence and absence of a given ligand). The cross-linking was performed in 0.2 M triethanolamine, pH 8.0. The reaction was started by the addition of the diimidoesters and allowed to proceed for 60 min at 30°C. The reaction was terminated by lowering the pH to 7.0 and the distribution of reaction products assessed by polyacrylamide gel electrophoresis in the presence of sodium dodecyl sulfate. The gels were stained with Coomassie Blue (the cross-linking did not change the staining properties of the protein). Panel A: C_d as a function of maximum effective reagent length (the maximal effective reagent lengths for malonic, adipic, pimelic, suberic, and dodecanedioic diimidates are 3.7, 7.3, 8.5, 9.7, and 14.5 A, respectively). (●), without AMP; (○), 0.1 mM AMP; (△), 0.3 mM AMP; and (□), 1.0 mM AMP. The points without AMP are the mean of at least four independent experiments. The lines connecting the points have no physical meaning; they only serve for better visualization. Panel B: r_K as a function of reagent length at various AMP concentrations. The symbols and number of experiments are the same as in panel A. (From Hajdu, J., Dombradi, V., Bot, G., and Friedrich, P., *Biochemistry*, 18, 4037, 1979. With permission.)

phenyl)dithio] propioimidate[19] (Figure 23). In these experiments, epidermal growth factor was reacted in the dark at pH 8.5 (0.1 M triethanolamine, 0.2 M NaCl, pH 8.5) for reaction of the imido ester formation with lysine. This reaction is terminated by the addition of ammonium acetate. Photoactivation in the presence of mouse 3T3 cells resulted in the specific labeling of a cell surface protein. It is noted that this reagent can be cleaved by reduction, permitting isolation of such a cell surface protein free of ligand.

The synthesis of *N*-(4-azidocarbonyl-3-hydroxyphenyl)-maleimide has been reported.[46] The reaction of this reagent with pig heart lactate dehydrogenase has been represented[46] and the site of modification has been established between a specific cysteine residue and a specific lysine residue.

The use of 4-nitrophenyl esters for cross-linking and affinity labeling has been proposed.[47]

The synthesis of 4-(6-formyl-3-azidophenoxy) butyrimidate (Figure 24) has been reported.[48] The imido function can react with the amino groups on a protein (0.1 M sodium borate, pH 8.6). Cross-linking can occur either with nitrene formation from the azido functional group upon irradiation or via reductive alkylation utilizing the aldehyde formation.

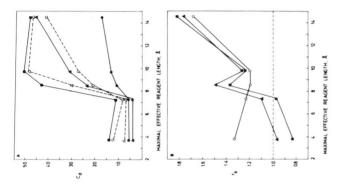

FIGURE 16. Structural changes in glycogen phosphorylase as revealed by cross-linking with bifunctional diimidates. Shown is the effect of phosphorylation and AMP on the cross-link parameters of phosphorylase. The experiments were performed as described in Figure 14. The cross-link parameters r_k and C_d are defined in the caption to Figure 14. Panel A shows a C_d distance diagram. Symbols: (●), phosphorylase b; (○), phosphorylase b plus 0.3 mM AMP; (▲), phosphorylase ab; (△), phosphorylase ab plus 0.3 mM AMP; and (■), phosphorylase a. The points are the mean of three independent experiments. Note that the abscissa is discontinuous; the lines connecting the points only serve for better visualization. Panel B: an r_k distance diagram. The reagent lengths, symbols, and abscissa are as in panel A. The r_k values were calculated by taking the rate constants measured with phosphorylase b as k_O. The points are the mean of three independent experiments. (From Dombradi, V., Hajdu, J., Bot, G., and Friedrich, P., *Biochemistry*, 19, 2295, 1980. With permission.)

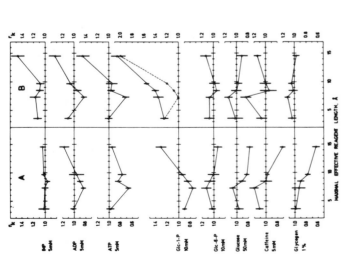

FIGURE 15. Structural changes in glycogen phosphorylase b as revealed by cross-linking with bifunctional diimidates. Shown is the effect of ligands on the r_k (see Figure 14) values of phosphorylase b cross-linking. The experiments were performed as described in Figure 14. The r_k values are shown as a function of maximal effective reagent length. Panel A shows cross-linking in the presence of a single ligand, as indicated. Panel B shows the cross-linking in the presence of 0.3 mM AMP + the ligand indicated on the left. The points connected by dotted lines for glucose-1-phosphate (Glc-1-P) + AMP denote the calculated average of the r_k values obtained with the two ligands when applied separately. The mean ± standard deviation of four independent experiments is illustrated. The lines connecting the points only serve for better visualization. (From Hajdu, J., Dombradi, V., Bot, G., and Friedrich, P., *Biochemistry*, 18, 4037, 1979. With permission.)

FIGURE 17. Structural changes in glycogen phosphorylase as revealed by cross-linking with bifunctional diimidates. Shown is the effect of ligands on the r_k cross-link parameter (defined in legend to Figure 14). Panel A represents studies performed with phosphorylase ab and panel B represents studies performed with phosphorylase a. The experimental conditions are described in Figure 14 except that the cross-linking was carried out at 18°C in the experiments described in panel B. The abscissa is discontinuous. The mean ± standard deviation of three independent experiments is shown. (From Dombradi, V., Hajdu, J., Bot, G., and Friedrich, P., *Biochemistry,* 19, 2295, 1980. With permission.)

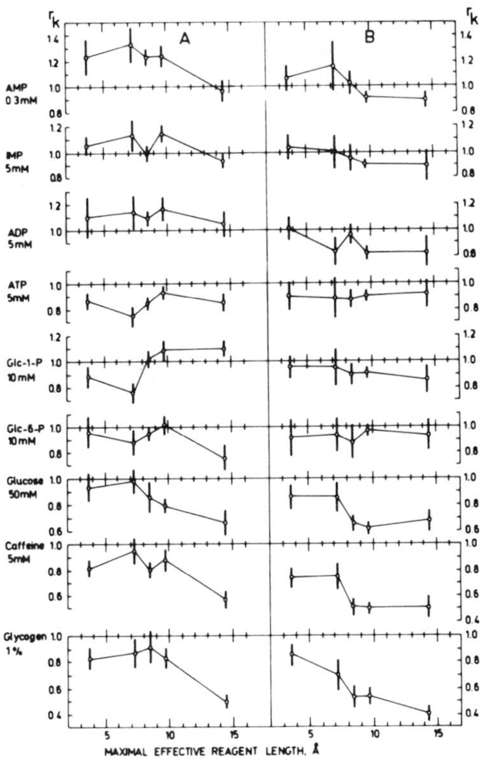

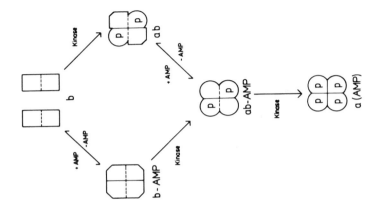

FIGURE 18. Structural changes in glycogen phosphorylase as revealed by cross-linking with bifunctional diimidates. Shown are the structural states detectable by cross-linking in rabbit muscle phosphorylase. Letters a, b, and ab denoted the respective forms of phosphorylase. Square, truncated square, and circular protomer symbols indicated by the dotted line denote allosterically competent subunit contact m. "Kinase" means the incorporation of phosphate (P) at Ser-14 by phosphorylase kinase. (From Dombradi, V., Hajdu, J., Bot, G., and Freidrich, P., *Biochemistry,* 19, 2295, 1980. With permission.)

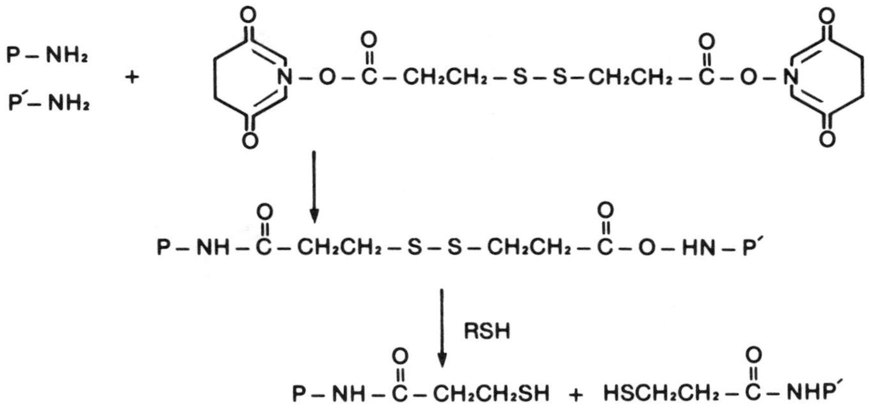

FIGURE 19.   The structure of 3,3'-dithiobis(succinimidylpropionate) and the formation of reversible cross-links in proteins via coupling with primary amino groups.

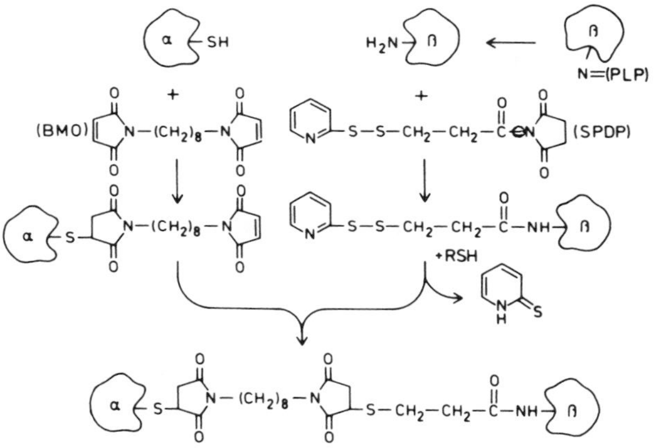

FIGURE 20.   The use of bifunctional maleimides to determine the solution structure of tryptophan synthetase. Shown is a scheme for the reactions leading to covalent cross-linking of α and β subunits. β Signifies one subunit within the β_2 dimer after modification by methyl acetimidate of the bulk of the NH$_2$ groups accessible in the holo enzyme; (PLP) is the pyridoxal phosphate group covalently bound to β by Schiff base formation. (BMO) is bis-*N*-maleimideo-1,8-octane; (SPDP) is *N*-succinimidyl-3-(2-pyridyldithio) propionate. (From Heilmann, H. D. and Holzner, M., *Biochem. Biophys. Res. Commun.*, 99, 1146, 1981. With permission.)

In either instance, radiolabel can be introduced with sodium borotritiide (NaB[^{3}T]$_4$) with conversion of the free aldehyde to an alcohol or by reducing the Schiff base formed during reductive alkylation.

The synthesis of *p*-azidophenylglyoxal (Figure 25) has been reported.[49] In the dark, this reagent shows a specificity similar to that for phenylglyoxal[49,50] with reaction primarily at arginine residues. Cross-linking is achieved by photoactivation of the azido function to the corresponding nitrene.

Chong and Hodges[51,52] have recently reported studies of a complex bifunctional affinity reagent, *N*-(4-azidobenzoylglycyl)-5-(2-thiopyridyl)-cysteine (AGTC) (Figure 26). Figure

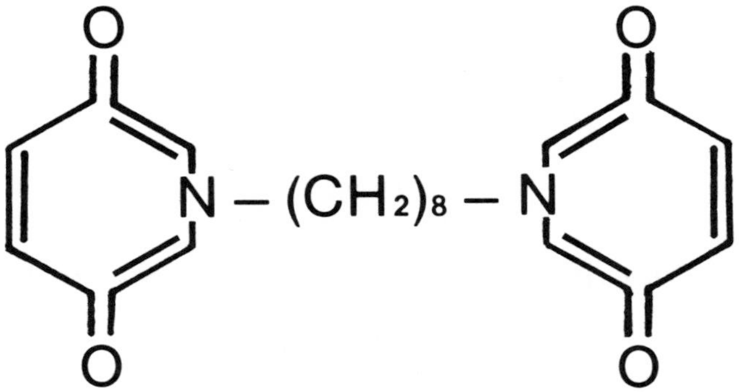

FIGURE 21. The structure of bis-*N*-maleimido-1,8-octane.

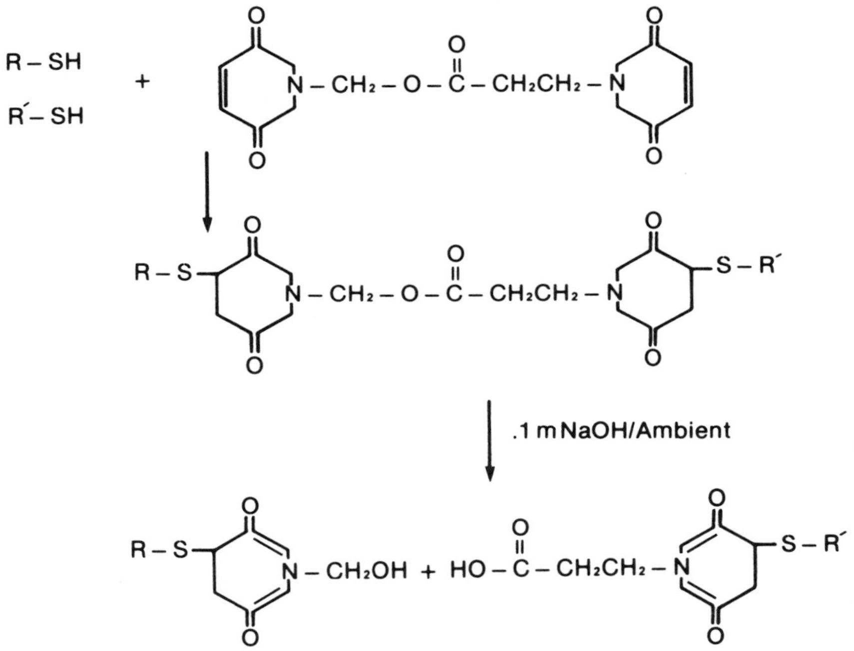

FIGURE 22. The structure of maleimidomethyl-3-maleimido propionate, the reaction of this reagent with protein sulfhydryl groups and the subsequent cleavage of this cross-link with base.

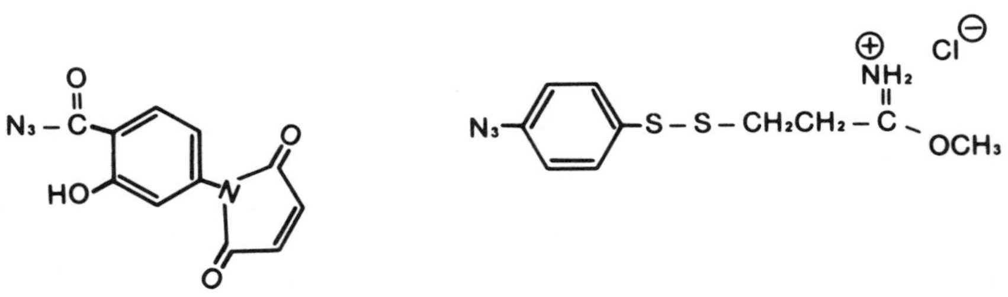

FIGURE 23. The structure of *N*-(4-azidocarbonyl-3-hydroxyphenyl)maleimide (on the left) and methyl-3-(*p*-azidophenyl)dithiopropioimidate (on the right).

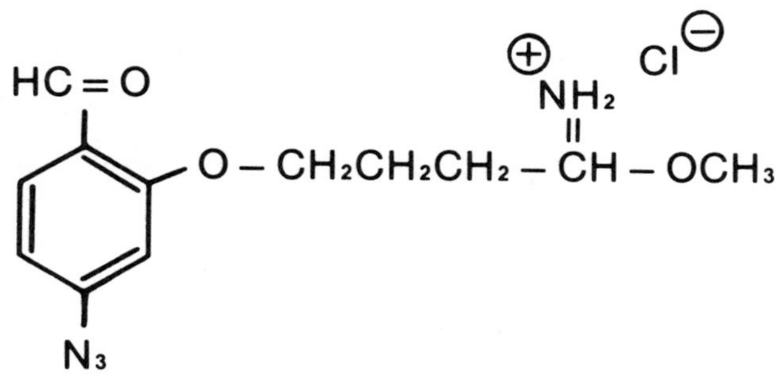

FIGURE 24.   The structure of 4-(6-formyl-3-azidophenoxy)butyrimidate.

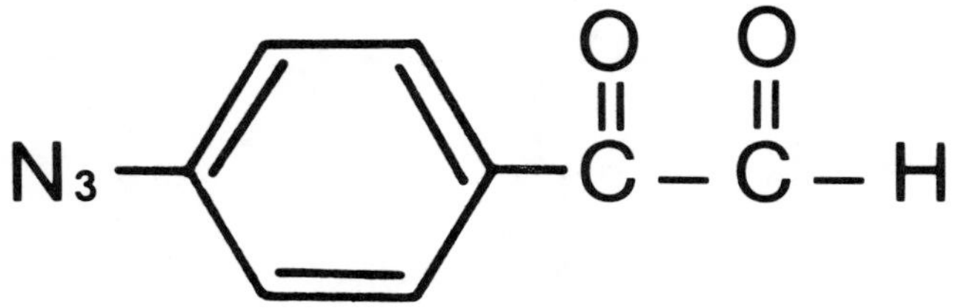

FIGURE 25.   The structure of *p*-azidophenylglyoxal.

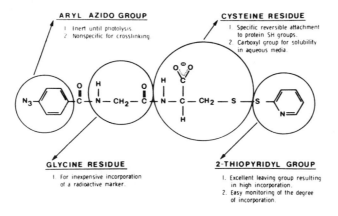

FIGURE 26.   The design of a heterobifunctional cross-linking reagent for the study of biological interactions between proteins. Shown is the rationale for the design of *N*-(4-azidobenzoylglycyl)-*S*-(2-thiopyridyl)-cysteine(AGTC). (From Chong, P.C.S. and Hodges, R. S., *J. Biol. Chem.*, 256, 5064, 1981. With permission.)

27 describes the effect of photolysis (350 nm) and reduction (dithiothreitol) on the spectral properties of the reagent. Figure 28 describes strategy in the use of this reagent.

One of the uses of cross-linking reagents such as those described above is the characterization of the interactions(s) between biologically active peptides and proteins and cell surface receptors. An excellent example of this approach is provided by the work of Ji and co-workers.[20,53] Some reagents developed for this purpose[54] are shown in Figure 29.

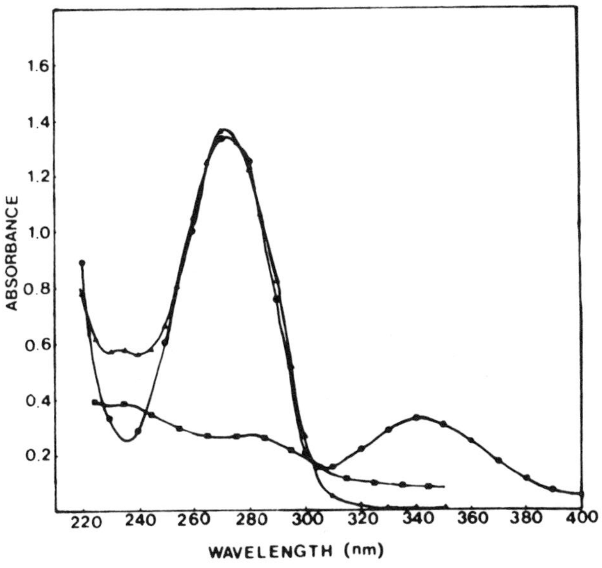

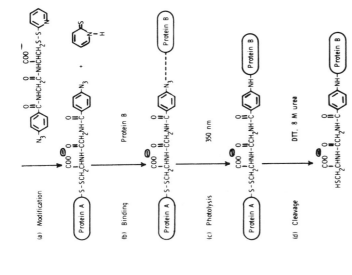

FIGURE 28.   The use of a heterobifunctional cross-linking reagent for the study of the biological interactions between proteins. Shown is a scheme for the radiolabeling of a binding site on one protein which is in the vicinity of a SH group on another protein. The SH groups are modified with the heterobifunctional photoaffinity probe N-(4-azidobenzoyl[2-³H]glycyl)-S-(2-thiopyridyl)cysteine. a, the introduction of the arylazide structure into a thioprotein A by thiol-disulfide interchange; b, noncovalent binding of modified protein A to protein B; c, covalent cross-linking of the proteins by photolysis of the arylazide moiety; and d, the cleavage of the disulfide bridge linking the two proteins by dithiothreitol, thus completing the transfer of radiolabel from protein A to protein B. (From Chong, P.C.S. and Hodges, R. S., *J. Biol. Chem.*, 256, 5071, 1981. With permission.)

FIGURE 27.   The effect of photolysis and dithiothreitol on the UV absorption spectrum of the heterobifunctional photoaffinity probe AGTC (see Figure 26). The spectrum of AGTC (50 μM in 50 mM Tris-HCl, 0.1 M KCl buffer, pH 7.5, before (▲——▲) and after either irradiation at 350 nm at 32°C for 30 min (■——■) or treatment with 25 mM dithiothreitol for 20 min (●——●). (From Chong, P.C.S. and Hodges, R. S., *J. Biol. Chem.*, 256, 5064, 1981. With permission.)

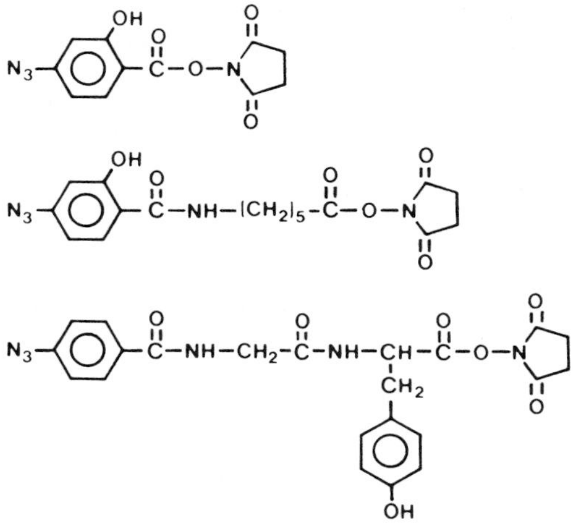

FIGURE 29. Structures of some cross-linking reagents which have proved useful in the study of the interaction of peptide hormones with cell surface receptors. Shown is NHS-ASA, the *N*-hydroxysuccinimide ester of 4-azido-salicylic acid; NHS-ABGT, the *N*-hydroxysuccinimide ester of 4-azidoben-zoylglycyltyrosine; and HNS-ASC, the *N*-hydroxysuccinimide ester of *N*-(4-azidosalicyl)-6-aminocaproic acid. (From Ji, T. H. and Ji, I., *Analyt. Biochem.*, 121, 286, 1982. With permission.)

REFERENCES

1. **Iwanij, V.,** The use of liver transglutaminase for protein labeling, *Eur. J. Biochem.*, 80, 359, 1977.
2. **Folk, J. E. and Finlayson, J. S.,** The ε-(γ-glutamyl) lysine crosslink and the catalytic rate of transglutaminases, *Adv. Protein Chem.*, 31, 1, 1977.
3. **Traub, W. and Piez, K. A.,** The chemistry and structure of collagen, *Adv. Protein Chem.*, 25, 243, 1971.
4. **Bornstein, P. and Sage, H.,** Structurally distinct collagen types, *Ann. Rev. Biochem.*, 49, 957, 1980.
5. **Husain, S. S. and Lowe, G.,** Evidence for histidine in the active sites of ficin and stem-bromelain, *Biochem. J.*, 110, 53, 1968.
6. **Moore, J., Jr. and Fenselau, A.,** Reaction of glyceraldehyde-3-phosphate dehydrogenase with dibromoacetone, *Biochemistry*, 11, 3753, 1972.
7. **Husain, S. S., Ferguson, J. B., and Fruton, J. S.,** Bifunctional inhibitors of pepsin, *Proc. Natl. Acad. Sci. U.S.A.*, 68, 2765, 1971.
8. **Lundblad, R. L. and Stein, W. H.,** On the reaction of diazoacetyl compounds with pepsin, *J. Biol. Chem.*, 244, 154, 1969.
9. **Mitra, S. and Lawton, R. G.,** Reagents for the cross-linking of proteins by equilibrium transfer alkylation, *J. Am. Chem. Soc.*, 101, 3097, 1979.
10. **Ruoho, A., Bartlett, P. A., Dutton, A., and Singer, S. J.,** A disulfide-bridge bifunctional imidoester as a reversible cross-linking reagent, *Biochem. Biophys. Res. Commun.*, 63, 417, 1975.
11. **Steele, J. C. H., Jr. and Nielson, T. B.,** Evidence of cross-linked polypeptides in SDS gel electrophoresis, *Analyt. Biochem.*, 84, 218, 1978.
12. **Davies, G. E. and Stark, G. R.,** Use of dimethyl suberimidate, a cross-linking reagent, in studying the subunit structure of oligomeric proteins, *Proc. Natl. Acad. Sci. U.S.A.*, 66, 651, 1970.
13. **Carpenter, F. H. and Harrington, K. T.,** Intermolecular cross-linking of monomeric proteins and cross-linking of oligomeric proteins as a probe of quaternary structure. Application to leucine aminopeptidase (bovine lens), *J. Biol. Chem.*, 247, 5580, 1972.
14. **Tarvers, R. C., Noyes, C. M., Roberts, H. R., and Lundblad, R. L.,** Influence of metal ions on prothrombin self-association. Demonstration of dimer formation by intermolecular cross-linking with dithiobis(succinimidyl propionate), *J. Biol. Chem.*, 257, 10708, 1982.

15. **Lewis, R. V., Roberts, M. F., Dennis, E. A., and Allison, W. S.,** Photoactivated heterobifunctional cross-linking reagents which demonstrate the aggregation state of phospholipase A$_2$, *Biochemistry,* 16, 5650, 1977.

16. **DeAbreu, R. A., DeVries, J., DeKok, A., and Veeger, C.,** Cross-linking studies with the pyruvate dehydrogenase complexes from *Azotobacter vinelandii* and *Escherichia coli, Eur. J. Biochem.,* 97, 379, 1979.

17. **Monneron, A. and d'Alayer, J.,** Effects of cross-linking agents on adenylate cyclase regulation, *FEBS Lett.,* 109, 75, 1980.

18. **Baskin, L. S. and Yang, C. S.,** Cross-linking studies of cytochrome P-450 and reduced nicotinamide adenine dinucleotide phosphate-cytochrome P-450 reductase, *Biochemistry,* 19, 2260, 1980.

19. **Das, M., Miyakawa, T., Fox, C. F., Pruss, R. M., Aharonov, A., and Herschman, H. R.,** Specific radiolabeling of a cell surface receptor for epidermal growth factor, *Proc. Natl. Acad. Sci. U.S.A.,* 74, 2790, 1977.

20. **Ji, T. H.,** A novel approach to the identification of surface receptors. The use of photosensitive hetero-bifunctional cross-linking reagent, *J. Biol. Chem.,* 252, 1566, 1977.

21. **Pilch, P. R. and Czech, M. P.,** Interaction of cross-linking agents with the insulin effector system of isolated fat cells. Covalent linkage of ^{125}I-insulin to a plasma membrane receptor protein of 140,000 daltons, *J. Biol. Chem.,* 254, 3375, 1979.

22. **Birnbaumer, M. F., Schrader, W. T., and O'Malley, B. W.,** Chemical cross-linking of chick oviduct progesterone-receptor subunits by using a reversible bifunctional cross-linking agent, *Biochem. J.,* 181, 201, 1979.

23. **Kasuga, M., Van Obberghen, E., Nissley, S. P., and Rechler, M. M.,** Demonstration of two subtypes of insulin-like growth factor receptors by affinity cross-linking, *J. Biol. Chem.,* 256, 5305, 1981.

24. **Rebois, R. V., Omedeo-Sale, F., Brady, R. O., and Fishman, P. H.,** Covalent cross-linking of human chorionic gonadotropin to its receptor in rat testes, *Proc. Natl. Acad. Sci. U.S.A.,* 78, 2086, 1981.

25. **Richards, F. M. and Knowles, J. R.,** Glutaraldehyde as a protein cross-linking reagent, *J. Mol. Biol.,* 37, 231, 1968.

26. **Monsan, P., Puzo, G., and Mazarguil, H.,** Etude du mecánisme d'établissement des liaisons glutaral-déhyde-protéins, *Biochimie,* 57, 1281, 1975.

27. **Quiocho, F. A. and Richards, F. M.,** Intermolecular cross-linking of a protein in the crystalline state: carboxypeptidase A, *Proc. Natl. Acad. Sci. U.S.A.,* 52, 833, 1964.

28. **Wong, C., Lee, T. J., Lee, T. Y., Lu, T. H., and Hung, C. S.,** Intermolecular cross-linking of a protein crystal - acid protease from *Endothia parasitica* — in 2.7 *M* ammonium sulfate solution, *Biochem. Biophys. Res. Commun.,* 80, 886, 1978.

29. **Spillburg, C. A., Bethune, J. L., and Vallee, B. L.,** Kinetic properties of crystalline enzymes. Carboxypeptidase A, *Biochemistry,* 16, 1142, 1977.

30. **Wong, C., Lee, T. J., Lee, T. Y., Lu, T. H., and Hung, C. S.,** The structure of acid protease from *Endothia parasitica* in cross-linked form at 3.5 Å resolution, *Biochem. Biophys. Res. Commun.,* 80, 891, 1978.

31. **Tüchsen, E., Hvidt, A., and Ottesen, M.,** Enzymes immobilized as crystals. Hydrogen isotope exchange of crystalline lysozyme, *Biochimie,* 62, 563, 1980.

32. **Habeeb, A. F. S. A. and Hiramoto, R.,** Reaction of proteins with glutaraldehyde, *Arch. Biochem. Biophys.,* 126, 16, 1968.

33. **Jansen, E. F., Tomimatsu, Y., and Olson, A. C.,** Cross-linking of α-chymotrypsin and other proteins by reaction with glutaraldehyde, *Arch. Biochem. Biophys.,* 144, 394, 1971.

34. **Hermann, R., Rudolph, R., and Jaenicke, R.,** Kinetics of *in vitro* reconstitution of oligomeric enzymes by cross-linking, *Nature (London),* 277, 243, 1979.

35. **Dutton, A., Adams, M., and Singer, S. J.,** Bifunctional imidoesters as cross-linking reagents, *Biochem. Biophys. Res. Commun.,* 23, 730, 1966.

36. **Monneron, A. and d'Alayer, J.,** Effects of imido-esters on membrane-bound adenylate cyclase, *FEBS Lett.,* 122, 241, 1980.

37. **Sinha, S. K. and Brew, K.,** A label selection procedure for determining the location of protein-protein interaction sites by cross-linking with bisimidoesters. Application to lactose synthase, *J. Biol. Chem.,* 256, 4193, 1981.

38. **Hajdu, J., Dombradi, V., Bot, G., and Friedrich, P.,** Structural changes in glycogen phosphorylase as revealed by cross-linking with bifunctional diimidates: phosphorylase b, *Biochemistry,* 18, 4037, 1979.

39. **Dombradi, V., Hajdu, J., Bot, G., and Friedrich, P.,** Structural changes in glycogen phosphorylase as revealed by cross-linking with bifunctional diimidates. Phospho-dephospho hybrid and phosphorylase a, *Biochemistry,* 19, 2295, 1980.

40. **Lomant, A. J. and Fairbanks, G.,** Chemical probes of extended biological structures: synthesis and properties of the cleavable protein cross-linking reagent [35S] dithiobis(succinimidyl propionate), *J. Mol. Biol.,* 104, 243, 1976.

41. **Lutz, H. U., Lomant, A. J., McMillan, P., and Wehrli, F.,** Rearrangement of integral membrane components during in vitro aging of sheep erythrocyte membranes, *J. Cell. Biol.,* 74, 389, 1977.

42. **Royer, G. P., Ikeda, S., and Aso, K.,** Cross-linking of reversibly immobilized enzymes, *FEBS Lett.,* 80, 89, 1977.

43. **Tarvers, R., Roberts, H. R., and Lundblad, R. L.,** Self association of bovine prothrombin fragment 1 in the presence of metal ions. The use of a covalent cross-linking reagent to study the reaction, *J. Biol. Chem.,* in press.

44. **Heilmann, H. D. and Holzner, M.,** The spatial organization of the active sites of the bifunctional oligomeric enzyme tryptophan synthetase: cross-linking by a novel method, *Biochem. Biophys. Res. Commun.,* 99, 1146, 1981.

45. **Sato, S. and Nakao, M.,** Cross-linking of intact erythrocyte membrane with a newly synthesized cleavable bifunctional reagent, *J. Biochem.,* 90, 1177, 1981.

46. **Kolkenbrock, H., Kiltz, H.-H., and Trommer, W. E.,** Stepwise cross-linking of pig heart lactate dehydrogenase by a heterobifunctional reagent, *Biochem. Biophys. Acta,* 535, 216, 1978.

47. **Jelenc, P. C., Cantor, C. R., and Simon, S. R.,** High yield photoreagents for protein cross-linking and affinity labeling, *Proc. Natl. Acad. Sci. U.S.A.,* 75, 3564, 1978.

48. **Maassen, J. A.,** Cross-linking of ribosomal proteins by 4-(6-formyl-3-azidophenoxy) butyrimidate. A heterobifunctional cleavable cross-linker, *Biochemistry,* 18, 1288, 1979.

49. **Ngo, T. T., Yam, C. F., Lenhoff, H. M., and Ivy, J.,** p-Azidophenylglyoxal. A heterobifunctional photoactivable cross-linking reagent selective for arginyl residues, *J. Biol. Chem.,* 256, 11313, 1981.

50. **Takahashi, K.,** The reaction of phenylglyoxal with arginine residues in proteins, *J. Biol. Chem.,* 243, 6171, 1968.

51. **Chong, P. C. S. and Hodges, R. S.,** A new heterobifunctional cross-linking reagent for the study of biological interactions between proteins. I. Design, synthesis and characterization, *J. Biol. Chem.,* 256, 5064, 1981.

52. **Chong, P. C. S. and Hodges, R. S.,** A new heterobifunctional cross-linking reagent for the study of biological interactions between proteins. II. Application to the troponin C-troponin I interaction, *J. Biol. Chem.,* 256, 5071, 1981.

53. **Ji, I., Yoo, B. Y., Kaltenbach, C., and Ji, T. H.,** Structure of the lutropin receptor on granulosa cells. Photoaffinity labeling with the α-subunit in human choriogonadotropin, *J. Biol. Chem.,* 256, 10853, 1981.

54. **Ji, T. H. and Ji, I.,** Macromolecular photoaffinity labeling with radioactive photoactivable heterobifunctional reagents, *Analyt. Biochem.,* 121, 286, 1982.

Chapter 6

AFFINITY LABELING

The use of biological affinity to label amino acid residues at enzyme active sites, allosteric binding sites, substrate binding sites, and other types of binding sites on proteins (such as the sites of binding of fatty acids and various other compounds to albumin) has proved to be a powerful tool in the study of the relationship between structure and function. This is an area of extensive study and it would be difficult to discuss all investigations in this area. The reader is directed to several recent comprehensive reviews in this area[1,2] including the recent review by Plapp[3] concerned with the application of affinity labeling to enzymes.

The concept of affinity labeling can be said to have been first advanced by Singer and co-workers.[4] The early work was expanded by the studies of Baker[5] which covered the early years of this area of investigation. It is, however, fair to suggest that studies of Shaw and co-workers on the labeling of the active-site histidine residues with peptide chloromethyl-ketones[6-8] really initiated serious study in this area. Details of these studies are presented below.

Affinity labeling is a technique for the specific modification of an amino acid residue in a protein which involves both the binding of the reagent (affinity label) on the basis of biological specificity and subsequent modification of an amino acid residue through the formation of a covalent bond. There are therefore at least two separate and distinct steps in the process of affinity labeling regardless of whether one is concerned with modification of an enzyme catalytic site or a binding site on a protein not involved in catalysis: the process of specific (selective) binding and the process of covalent bond formation.

It therefore follows that the process of affinity labeling should show saturation kinetics such as that shown for the reaction of human placental 17β-estradiol dehydrogenase/20α-hydroxysteroid dehydrogenase[9] with 5'-p-fluorosulfonyl adenosine (Figure 1). The kinetics of modification can be demonstrated as:

$$P + AL \underset{k_2}{\overset{k_1}{\rightleftharpoons}} P \cdot AL \overset{k_3}{\rightarrow} P-AL \tag{1}$$

$$K_{AL} = \frac{[P][AL]}{[P \cdot AL]}; \quad k_{observed} = \frac{k_3[AL]}{K_{AL} + [AL]} \tag{2}$$

$$\frac{1}{k_{observed}} = \frac{K_{AL}}{k_3[AL]} + \frac{1}{k_3} \tag{3}$$

where P is the protein (not necessarily an enzyme), AL is the affinity label, P·AL is the noncovalent complex, and P−AL is the covalently bonded product of the reaction. This approach has been adapted from Powers and co-workers[10] and implies that when reagent (AL) concentration is much greater than protein concentration that decrease of P + P·AL in the reaction mixture will follow pseudo first-order kinetics at a fixed value of AL. The observed reaction rate ($k_{observed}$) is not constant with changes in the concentration of affinity label. However, as noted by these investigators, $k_{observed}/[AL]$ remains constant over a substantial concentration range for peptide chloromethyl ketones and α-chymotrypsin (Table 1).[10] This is explained by $k_{AL} \gg [AL]$ such that:

$$\frac{k_{observed}}{[AL]} = \frac{k_3}{K_{AL}} \tag{4}$$

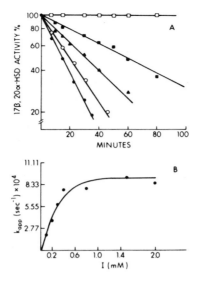

FIGURE 1.   Saturation kinetics in the reaction of 5′-[*p*-(fluorosulfonyl)benzoyl]adenosine with 17β-estradiol/20α-hydroxysteroid dehydrogenase. Shown is the inactivation of the 17β-estradiol dehydrogenase and 20α-hydroxysteroid dehydrogenase activities by various concentrations of 5′-[*p*-(fluorosulfonyl)benzoyl]adenosine. In panel A the enzyme (2μ*M*) in 6 mℓ 0.01 *M* potassium phosphate, pH 7.0, containing 5 m*M* EDTA and 20% (V/V) glycerol was incubated at 25°C with inactivator [final concentrations (■) 100 μ*M*; (▲) 200 μ*M*; (○) 300 μ*M*, and (●) 400 μ*M*] in 0.12 mℓ of ethylene glycol. Identical control incubations contained adenosine. At the indicated times, portions were removed and assayed for both 17β and 20α activities. The percentage of enzyme activity is a logarithmic scale along the ordinate. For simplification of the graphic presentation, the single points represent the mean of duplicate assays for both activities. In panel B the data from panel A as well as inactivation studies using saturating concentrations of inactivator (500 to 2000 μ*M*) were used to calculate the k_{app} (0.693/$t_{1/2}$) and to construct the plot demonstrating saturation kinetics. (From Tobias, B. and Strickler, R. C., *Biochemistry*, 20, 5546, 1981. With permission.)

Table 1
REACTION OF PEPTIDE
CHLOROMETHYL KETONES
WITH α-CHYMOTRYPSIN[10]

Inhibitor	$k_{observed}$/[AL]
N-Formyl-PheCH$_2$Cl	0.05
N-Acetyl-PheCH$_2$Cl	0.15
N-Tosyl-PheCH$_2$Cl	0.12
N-Acetyl-GlyPheCH$_2$Cl	0.13
N-Acetyl-AlaPheCH$_2$Cl	0.11
N-Acetyl-LeuPheCH$_2$Cl	0.32
N-Cb$_3$-GlyGlyPheCH$_2$Cl	0.32
N-Boc-AlaGlyPheCH$_2$Cl	0.33
N-Acetyl-AlaGlyPheCH$_2$Cl	0.88
N-Boc-GlyLeuPheCH$_2$Cl	1.29

The reader is referred to Powers and co-workers,[10,11] Wold,[12] and Plapp[3] for a more extensive consideration of the kinetics of interaction of affinity labels with proteins.

There are at least three major considerations which must be satisfied in order to justify

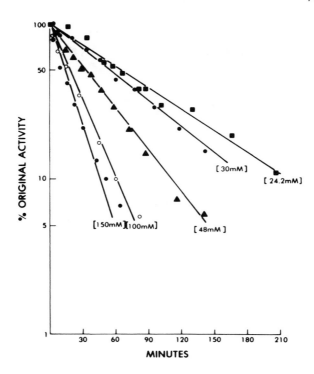

FIGURE 2. Loss of creatine kinase activity upon incubation with epoxycreatine (*N*-(2,3-epoxypropyl)-*N*–amidinoglycine). Conditions were as follows: creatine kinase (1 mg/mℓ) in 10 m*M* HEPES, pH 7.5, at 0°C. The epoxycreatine concentrations are noted on the figure. (From Marletta, M. A. and Kenyon, G. L., *J. Biol. Chem.*, 254, 1879, 1979. With permission.)

identification of a compound as a true affinity label. First, the reaction must show a rate saturation effect as one increases inhibitor (affinity label) concentration (Figure 1). Secondly, substrate, competitive inhibitor, or ligand must protect against modification and inactivation. Finally, the reaction must demonstrate stoichiometry with respect to sites modified/functional subunit. The modification of creatine kinase by *N*-(2,3-epoxypropyl)-*N*-amidinoglycine[13] provides an excellent example of these considerations. Figure 2 shows the time course for the loss of activity by creatine kinase upon reaction with the epoxycreatine derivative in 0.010 *M* HEPES, pH 7.5 (0°C) at several concentrations of reagent. When the $t^{1}/_{2}$ (inactivation halftime) is plotted vs. the reciprocal of inhibitor concentration a straight line is obtained[14] with the y intercept equal to $T^{1}/_{2}$ (minimum inactivation halftime). Such a graph for the data obtained from the experiments described in Figure 2 is shown in Figure 3. The minimum inactivation halftime obtained from this analysis is 4.2 min corresponding to a pseudo first-order rate constant of 2.8×10^{-3} sec^{-1}. A value for K_{INACT} (concentration of inhibitor giving half-maximal inactivation rate) is obtained either from the slope of the line or the negative of the reciprocal of the intercept on the X axis and, in these experiments, is 355 m*M*. The stoichiometry for this reaction is presented in Figure 4 and is consistent with the incorporation of 1 mol of reagent per mole of enzyme subunit. In the presence of epoxycreatine, there is an increase in amounts of ADP generated based on comparison with a control experiment in the absence of epoxycreatine suggesting that epoxycreatine is phosphorylated by the enzyme. This indicates that epoxycreatine does interact with the active site of the enzyme. Further support for this interaction is shown in Figure 5 where the Mg^{2+}-ADP-creatine complex does protect the enzyme from inactivation.

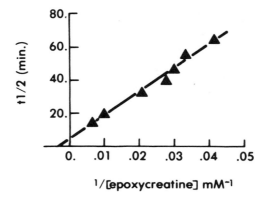

FIGURE 3.   The half-time of inactivation of creatine kinase as a function of the reciprocal of the epoxycreatine concentration. The half-times of inactivation were obtained from Figure 2. (From Marletta, M. A. and Kenyon, G. L., *J. Biol. Chem.*, 254, 1879, 1979. With permission.)

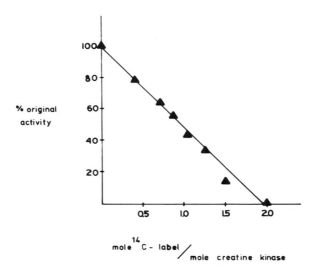

FIGURE 4.   Stoichiometry for the inactivation of creatine kinase by epoxycreatine. Shown is the loss of creatine kinase activity upon incorporation of [^{14}C]epoxycreatine. Conditions were as follows: creatine kinase (17 mg/mℓ, 209 μM) and epoxycreatine (39 mM) in 10 mM HEPES, pH 7.5 at 0°C. (From Marletta, M. A. and Kenyon, G. L., *J. Biol. Chem.*, 254, 1879, 1979. With permission.)

There is almost an infinite variety of compounds which can be used as affinity labels. First are derivatives which form relatively stable analogues/homologues of enzyme intermediates. Examples of this class include diisopropylphosphorofluoridate,[15] peptide aldehydes,[16] and compounds such as *p*-nitrophenyl-*p'*-guanidinobenzoate which serve as "active-site titrants" of serine proteases.[17-19] Second are the so-called K_s reagents, which are isosteric with respect to substrate or ligands and possess a relatively unreactive functional group such as a sulfonyl fluoride[20] (Figure 6) or a halomethyl ketone derivative such as L-1-tosylamido-

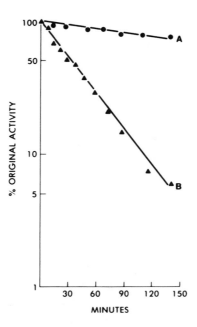

FIGURE 5. The ability of substrate to protect creatine kinase from inactivation by the active-site-directed reagent, *N*-(2,3-epoxypropyl)-*N*-amidinoglycine (epoxycreatine). Shown is protection from inactivation by the MgADP·NO$_3$ creatine complex. Conditions were as follows: (A) creatine kinase (1.0 mg/mℓ), epoxycreatine (48 m*M*), ADP (8 m*M*), magnesium acetate (8 m*M*), NaNO$_3$ (8 m*M*), and creatine (40 m*M*) in 10 m*M* HEPES, pH 7.5 at 0°C; (B) the same except ADP, magnesium acetate, NaNO$_3$, and creatine were all deleted. (From Marletta, M. A. and Kenyon, G. L., *J. Biol. Chem.*, 254, 1879, 1979. With permission.)

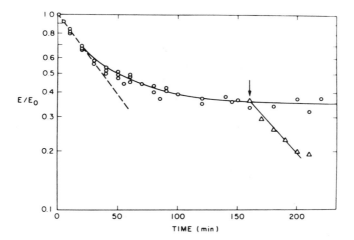

FIGURE 6. The inactivation of yeast pyruvate kinase by 5'-*p*-fluoro-sulfonylbenzoyl adenosine (5'-*p*-FSO$_2$BzAdo). Pyruvate kinase (0.22 to 0.49 mg/mℓ) was incubated with 1.1 m*M* 5'-*p*-FSO$_2$BzAdo (○ —— ○) at 25°C in 20 m*M* potassium barbital buffer, pH 8.6, containing 200 m*M* KCl, 0.5 m*M* EDTA, and 4.4% dimethylformamide. At the indicated times, a portion of the reaction mixture was withdrawn and assayed for enzymatic activity. After 160 min of reaction with the reagent, an addition 2 μℓ of 51 m*M* 5'-*p*-FSO$_2$BzAdo in dimethylformamide was added to 100 μℓ of the reaction mixture (△ —— △) and 2 μℓ of dimethylformamide to 100 μℓ of the control. (From Likos, J. J., Hess, B., and Colman, R. F., *J. Biol. Chem.*, 255, 9388, 1980. With permission.)

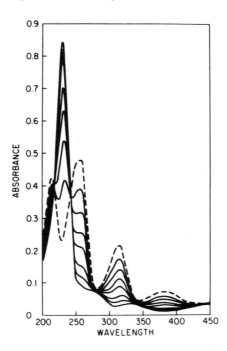

FIGURE 7.   Change in the UV absorption spectrum of 3-azido-2,7-naphthalene disulfonate (ANDS) upon irradiation. Shown are the flashed spectra of 30 μ*M* ANDS in 10 m*M* sodium phosphate, pH 7.0, at 25°C. The dashed line is the spectrum of the ANDS before flash. Each solid line represents a 15 sec exposure to a 366 nm hand-held UV light. The λmax of ANDS shifts from 260 nm to 233 nm during irradiation. (From Moreland, R. B. and Dockler, M. E., *Analyt. Biochem.*, 103, 26, 1980. With permission.)

2-phenylethyl chloromethyl ketone (TPCK),[6] bromoacetylcholine,[21] and 1-chloro-2-oxo-hexanol-6-phosphate.[22] With this type of reagent, the functional group (i.e., sulfonyl halide, α-keto alkylhalide) is comparatively unreactive and reaction with a nucleophile (i.e., lysine, cysteine, histidine) is accomplished by enhanced local reagent concentration caused by specific binding. In another type, reaction is accomplished by generation of a reactive species after binding. The most popular reagents in this category are aryl azides which, upon irradiation by UV light, will generate short-lived, highly reactive nitrenes. Examples of this class (type) of reagent include 4-fluoro-3-nitrophenyl azide,[24] *N*-4-azido-2-nitrophenyl-γ-aminobutyryl ADP[25] and 8-azidoadenosine 3′,5′-monophosphate[25] and 3-azido-2,7-naphthalene disulfonate[27] (Figure 7). There are also other compounds which are photosensitive in that irradiation generates a rapidly reacting species. Examples in this general category include *p*-dimethylaminobenzene diazonium fluoride,[28] 4-nitrophenyl-α-D-galactopyranoside,[29] and 17β-hydroxy-4,6-androstadien-3-one (Δ[6]-testosterone).[30] The final group of reagents to be discussed includes the ''suicide'' substrates, in which the reactive functional group is generated by catalysis at the active site via the normal reaction mechanism.[31,32] These reagents will frequently provide information regarding enzyme mechanism as well as functional groups at the enzyme active site.

It should be stressed that affinity labeling is conceptually a different process than ''fortuitous'' modification of a uniquely reactive residue at, for instance, the active site of an enzyme.[33-37]

An excellent example of the use of increasing structural information in the quality of the inhibition (inactivation) is the work on chymotrypsin initiated by Schoellmann and Shaw[6]

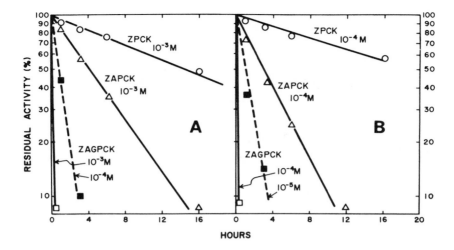

FIGURE 8. The inactivation of subtilisin BPN' by peptide chloromethyl ketone derivatives. The derivatives are: ZPCK, carbobenzoxy-L-phenylalanine chloromethyl ketone; ZAPCK, carbobenzoxy-L-phenylalanyl chloromethyl ketone; ZAGPCK, carbobenzoxy-L-alanyl-glycyl-L-phenylalanine chloromethyl ketone. Shown is the loss of esterase activity on incubation of the enzyme in 0.05 M Tris-HCl, pH 7.0, containing 0.001 M CaCl$_2$ and either 25% (panel A) or 10% (panel B) dioxane at 40°C. The enzyme concentration was 5 μM and inhibitor concentrations were as indicated in the figure. (From Morihara, K. and Oka, T., *Arch. Biochem. Biophys.*, 138, 526, 1970. With permission.)

and extended during the subsequent decade by Powers and co-workers. The initial studies used L-1-tosylamido-2-phenylethyl chloromethyl ketone.[6] Reaction of L-1-tosylamido-2-phenylethyl chloromethyl ketone (TosPheCH$_2$Cl) with chymotrypsin is dependent upon the structure of the native enzyme; chymotrypsin previously exposed to 8.0 M urea does not react with reagent. This incidentally is an excellent test for specificity of affinity labeling; reaction should not occur with denatured enzyme. The D-isomer of TosPheCH$_2$Cl does not react with the active site histidine.[38] These investigators also reported that modification of the methionine residue at the active site (Met-192) occurred, but this reaction is apparently nonspecific and stereochemistry of the modifying reagent did not have any apparent role. The methionine residue in question was modified by iodoacetate but not with the affinity label. Morihara and Oka[39] examined the reaction of chymotrypsin with a closely related compound, carbobenzoxy-L-phenylalanine chloromethyl ketone (CBZPheCH$_2$Cl). CBZPheCH$_2$Cl reacted effectively with the active site histidine of chymotrypsin; a greater than tenfold decrease in the observed first-order rate constant was seen if the complexity of the reagent was increased to carbobenzoxy-glycyl-L-phenylalanyl chloromethyl ketone. However if the complexity of the reagent was increased to carbobenzoxy-L-alanyl-glycyl·L-phenylalanine chloromethyl ketone, an approximate sevenfold increase in reaction rate was observed (Figure 8). It is of interest that CBZGlyPheCH$_2$Cl was approximately 8 times more effective than CBZPheCH$_2$Cl with subtilisin BPN' at pH 7.0 (0.05 M Tris, 0.001 M CaCl$_2$, 10% dioxane) while CBZAlaGlyPheCH$_2$Cl was greater than 200 times more effective. Thus, affinity labels need not show the same order of effectiveness with different enzymes of similar specificity.

As mentioned above, Powers and co-workers have pursued a systematic analysis of the reaction of peptide chloromethyl ketones with various serine proteases. Most noteworthy for present consideration is the study of the reaction of peptide chloromethyl ketones with chymotrypsin.[10] The results of a study of this influence of increasing structural quality of

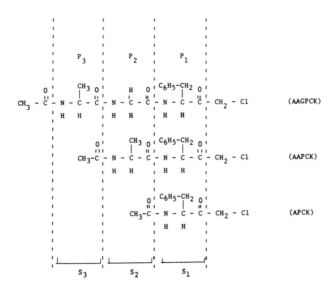

FIGURE 9.   Schematic diagram of the inhibitors, acetyl-L-alanylgly-
cyl-L-phenylalanine chloromethyl ketone (AAGPCK), acetyl-L-alanyl-
L-phenylalanine chloromethyl ketone (AAPCK) and acetyl-L-phenyl-
alanine chloromethyl ketone (APCK) and their interactions with the
binding site on chymotrypsin according to the notation of Schechter
and Berger. (From Segal, D. M. et al., *Cold Spring Harbor Symp.
Quant. Biol.*, 36, 85, 1971.

the inhibitor on the rate of inactivation of α-chymotrypsin is presented in Table 1. These
observations when combined with an elegant application of crystallographic analysis[40-42] (see
Figures 9, 10, and 11) provide considerable information regarding secondary substrate
binding sites. The concept of greatest importance in considering this information is that the
greater the specificity information (quality) built into an affinity label, the greater the significance of the study of the reaction of the compound with the protein (enzyme) under
investigation.

The reaction of peptide chloromethyl ketones with sulfhydryl proteases has been investigated.[43,44] Despite the presence of a histidine residue at the active site of these enzymes,
reaction occurs at the active site sulfhydryl group. Comparison of the rates of reaction of
TosLysCH$_2$Cl with free cysteine and with papain demonstrated that reaction with the enzyme
occurs approximately 10^6 more rapidly. Rate enhancement of this order of magnitude is seen
on occasion with putative affinity labels. While not necessarily as dramatic as this, differences
should be seen in comparing either the rates of reaction of affinity label with model compound
and protein (as seen above with papain) or the rates of reaction of affinity label and its
functional group with protein (see Reference 45 for a comparison of the rates of reaction of
N-bromoacetylglucosamine and bromoacetic acid with rat muscle hexokinase). The differences in the reaction of peptide chloromethyl ketones with serine proteases and sulfhydryl
proteases have been considered in some detail by Brocklehurst and Malthouse.[46]

As described above, α-keto-halo compounds can modify histidyl and/or cysteinyl residues
at the active site. Further versatility (variety) in the reaction of this class of derivative was
shown by the work of Hass and Neurath[47,48] on the affinity labeling of bovine carboxypeptidase A (Figures 12 and 13). Here the affinity label, N-bromoacetyl-N-methyl-L-phenyl-
alanine, reacts with the active site carboxylic acid (Glu-270) with the formation of a glycolic
acid derivative. It is of interest to note that the reaction of iodoacetate with ribonuclease T$_1$

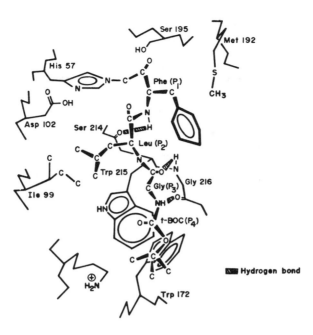

FIGURE 10. A schematic drawing of the inhibitor moiety, Boc-Gly-Leu-PheCH₂Cl, bound to chymotrypsin A$_\alpha$. Only the portion of the enzyme which can interact with the inhibitor is shown. (From Kurachi, K., Powers, J. C., and Wilcox, P. E., *Biochemistry,* 12, 771, 1973. With permission.)

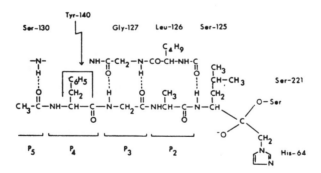

FIGURE 11. A schematic representation of the inhibitor moiety, Ac-Phe-Gly-Ala-LeuCH₂ bound to the active site of subtilisin BPN'. The probable interactions are based on crystallographic studies of subtilisin inhibited with other chloromethyl ketones. (From Powers, J.C., Lively, M.O.,III, and Tippett, J.T., *Biochim. Biophys. Acta,* 480, 246, 1977. With permission.)

as described by Takahashi and co-workers[49] results in the loss of enzyme activity concomitant with the modification of a glutamic acid residue at the active site. *N*-Bromoacetyl-L-phenylalanine is also an inhibitor but also serves as a substrate; *N*-methylation eliminates its ability to serve as a substrate. Modification of an active site carboxyl group with an α-keto-halo compound also occurs with an affinity label described by Rasnick and Powers[50] (Figures 14 and 15). The active site-directed inhibitor used, *N*-chloroacetyl-*N*-hydroxyleucine methyl

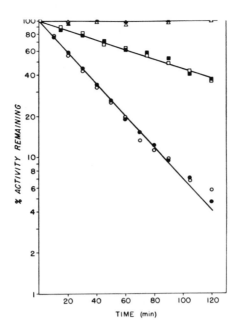

FIGURE 12.   Affinity labeling of bovine carboxypeptidase A_γ^{Leu} (CPA_γ^{Leu}) by *N*-bromo-acetyl-*N*-methyl-L-phenylalanine (BAMP). Shown is the loss of peptidase and esterase activities of CPA_γ^{Leu} as a function of time during incubation with L-BAMP, D-BAMP, and bromoacetate. CPA_γ^{leu} (0.1 mg/mℓ) was treated with 1 m*M* L-BAMP, 10 m*M* D-BAMP, or 10 m*M* bromoacetate in a solution containing 1 *N* NaCl, 0.05 *M* Tris-chloride (pH 7.5). The esterase (L-BAMP (●); D-BAMP (■); bromoacetate (▲)) and peptidase (L-BAMP (○) D-BAMP (□); bromoacetate (△)) activities of the enzyme were monitored as a function of time. (From Hass, G. M. and Neurath, H., *Biochemistry*, 10, 3535, 1971. With permission.)

ester, is novel in that a substantial portion of the "affinity" is provided by interaction with the zinc atom at the active site (see Figure 16). Replacement of the P′ amide hydroxyl with a hydrogen obviates the majority of the quality of this inhibitor.

Active-site directed inhibition can also be designed to probe regions surrounding the active site or primary substrate binding site (S_1 site; see Reference 51 for a discussion of this nomenclature which is essential for the understanding of secondary substrate binding sites). An example of this approach is found in the studies of Bing and co-workers[53,54] with m[*O*-(2-chloro-5-fluorosulfonylphenylureido)phenoxybutoxy] benzamidine (see Figure 17).

The development of peptide chloromethyl ketones which serve as affinity labels for serine protease-like enzymes with "tryptic"-like specificity serves as a good example of the logical development of an active-site directed inhibitor. It was first established that trypsin preferentially hydrolyzed peptide bonds in which the carboxyl group was provided by a lysine or arginine residue.[55] It was subsequently shown that various amides and guanidines would effectively bind to trypsin.[7] Inagami then demonstrated that alkylation of the active site histidine with iodoacetamide occurred much more rapidly in the presence of a reagent (i.e., methyl guanidine) than in the absence of such a compound.[56] These observations then led to the development of *N*-tosyl-L-lysine chloromethyl ketone (1-chloro-3-tosylamido-7-amino-2-heptanone, TLCK, TosLysCH₂Cl) as an active-site directed reagent (affinity label) (Figures 18 to 21) for trypsin[8] which was later shown to react with histidine residue 43 (His 57 in chymotrypsin numbering system) at the N-3 position.[58] It is of interest to note that another compound, *p*-guanidinophenacyl bromide, which is also an analogue of a trypsin substrate

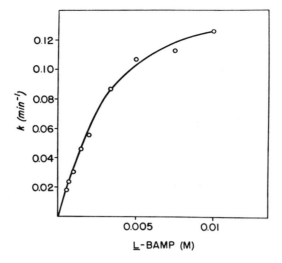

FIGURE 13. Psuedo first-order rate constant, k, for the inactivation of CPA_γ^{Leu} as a function of L-BAMP concentration at pH 7.5. Incubation mixtures contained 0.1 mg/mℓ of CPA_γ^{Leu} in 1 N NaCl, 0.05 M Tris-chloride (pH 7.5), including different amounts of L-BAMP. Esterase activity was monitored as a function of time at each concentration of L-BAMP and pseudo first-order rate constants were calculated from the following equation:

$$\ln\left(\frac{E_1}{E_2}\right) = k(T_2 - T_1)$$

(From Hass, G. M. and Neurath, H., *Biochemistry*, 10, 3535, 1971. With permission.)

IRREVERSIBLE INHIBITION OF THERMOLYSIN

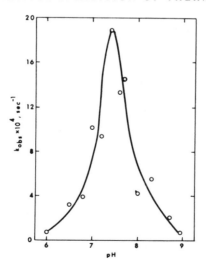

FIGURE 14. pH Dependence of the inactivation of thermolysin by *N*-chloroacetyl-DL-*N*-hydroxyleucine methyl ester ($ClCH_2CO$-DL-(N-OH)Leu-OCH_3). The concentration of the enzyme was 4.2 μM and the concentration of the inhibitor was 2.3 mM. The buffers used were: PIPES (pH 6.0), HEPES (pH 6.50 to 7.70), and Tris (pH 8.00 to 8.95); all were 0.1 M and contained 0.01 M $CaCl_2$. (From Rasnik, D. and Powers, J. C., *Biochemistry*, 17, 4363, 1978. With permission.)

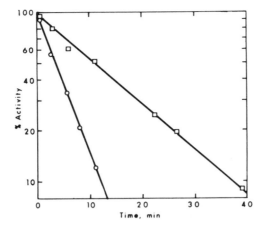

FIGURE 15.   Protection of thermolysin from inactivation by an active site-directed reagent, *N*-chloroacetyl-DL-*N*-hydroxyleucine methyl ester. Thermolysin (4.4 μ*M*) in 0.1 *M* Tris, 0.01 *M* CaCl$_2$, pH 7.2, containing 5% dimethylformamide was incubated with the reagent (4.5 m*M* in the presence (□) and absence (○) of the competitive inhibitor carbobenzoxy-L-phenylalanine (1.7 m*M*). (From Rasnick, D. and Powers, J. C., *Biochemistry*, 17, 4363, 1978. With permission.)

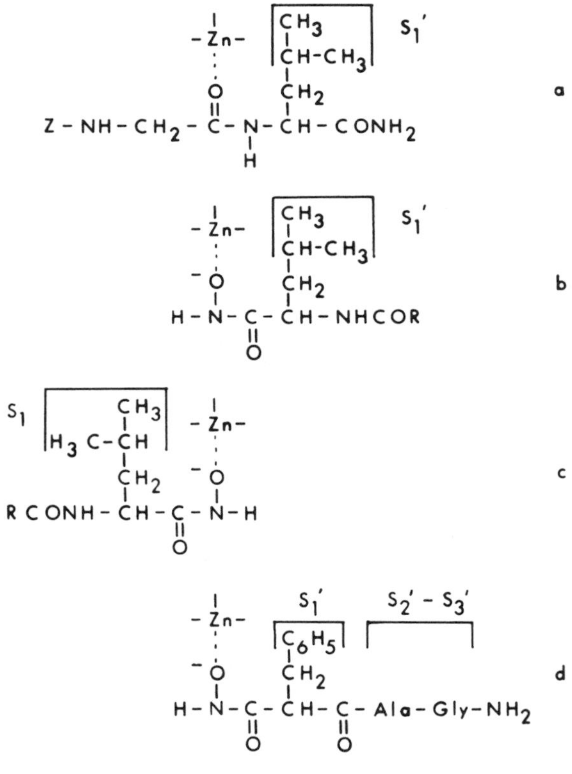

FIGURE 16.   Peptide hydroxamic acids as inhibitors of thermolysin. Shown are schematic drawings demonstrating the binding of a substrate (a) to the active site of thermolysin and possible binding modes for hydroxamic acid inhibitors (b-d). (From Nishino, N. and Powers, J. C., *Biochemistry*, 17, 2846, 1978. With permission.)

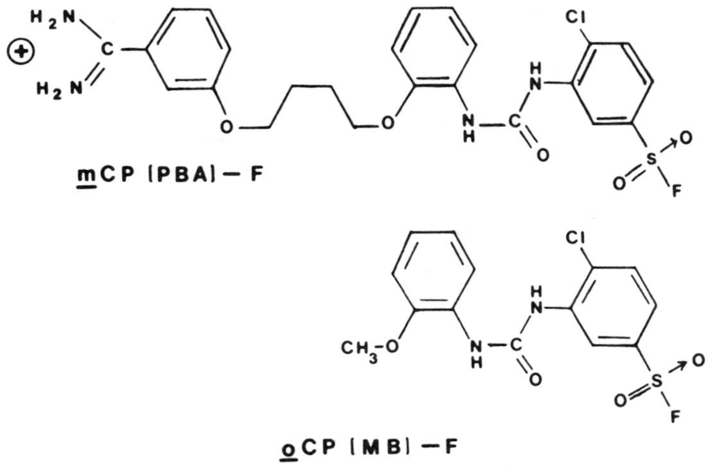

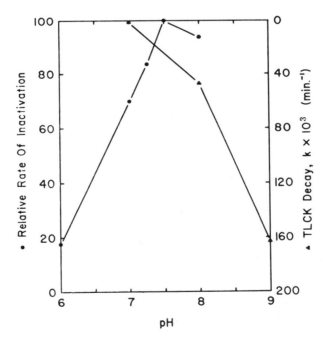

mCP (PBA) — F

oCP (MB) — F

FIGURE 17. The structures of certain "exo-site" affinity labels. Shown are the schematic structures of mCP(PBA)-F, *m*-[*o*-(2-chloro-5-fluorosulfonylphenylureido)phenoxybutoxy]benzamidine, and *p*-CP-(MB)-F, *o*-(2-chloro-5-fluorosulfonylphenylureido)methoxybenzene. The butyl connecting chain in *m*CP(PBA)-F is represented by a bent line with the -CH$_2$- groups corresponding to bends. (From Bing, D. H., Cory, M., and Fenton, J. W., II, *J. Biol. Chem.*, 252, 8027, 1977. With permission.)

FIGURE 18. The effect of pH on the reactivity of TLCK(1-chloro-3-tosylamido-7-amino-2-heptanone, also tosyllysyl chloromethyl ketone) with trypsin and on TLCK stability. The studies were performed in 0.2 *M* Tris-maleate buffers at the indicated pH. The values for the rate of inactivation of trypsin are not corrected for decomposition of TLCK. (From Shaw, E., Mares-Guia, M, and Cohen, W., *Biochemistry*, 4, 2219, 1965. With permission.)

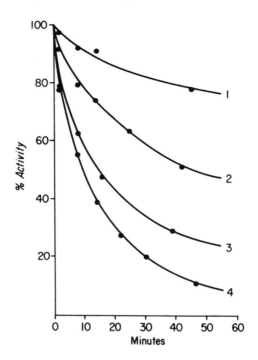

FIGURE 19.   The protection of trypsin from inactivation by 1-chloro-3-tosylamido-7-amino-2-heptanone
(TLCK). Shown is the protection of trypsin against inactivation by TLCK with benzamidine as a reversible
inhibitor. The concentration of trypsin was 45.6 μM and
the concentration of TLCK was 0.617 mM. The experiments were performed in 0.02 M Veronal, pH 8.0 at
25°C. Curve 1; benzamidine, 0.783 mM; curve 2, benzamidine, 0.188 mM; curve 3, benzamidine, 0.0783
mM; and curve 4, no benzamidine. (From Shaw, E.,
Mares-Guia, M., and Cohen, W., *Biochemistry*, 4, 2219,
1965. With permission.)

(inhibitor) inactivates trypsin by modification of the active site serine residue (Ser-183)[59]
(see Figure 22).

Photoaffinity labels have become extremely popular. These reagents are essentially unreactive until exposed to light, whereupon the active species is generated (i.e., a nitrene is
formed from an azide) which then reacts instantaneously with a variety of groups. Particular
attention must be paid to photolysis rate and solvent conditions. As noted by DeTraglia and
co-workers,[60] characterization of the rate(s) of photolysis vs. wavelength of irradiation
provides useful information to control photoaffinity labeling (see Figures 23 and 24). Another
potential problem in the use of aryl azides is the ease of reduction to the corresponding
amine with a mild reducing agent such as dithiothreitol.[61] Selected examples of affinity
labels are given in Table 2.

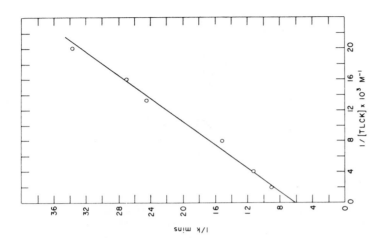

FIGURE 21. Demonstration of saturation kinetics in the inactivation of trypsin by 1-chloro-3-tosylamido-7-amino-2-heptanone (TLCK). The k_{app} of inactivation at various TLCK concentrations was obtained from experiments which provide first-order inactivation kinetics (see Figure 20). Shown is a double-reciprocal plot of k_{app} vs. TLCK concentration. (From Shaw, E. and Glover, G., *Arch. Biochem. Biophys.*, 139, 298, 1970. With permission.)

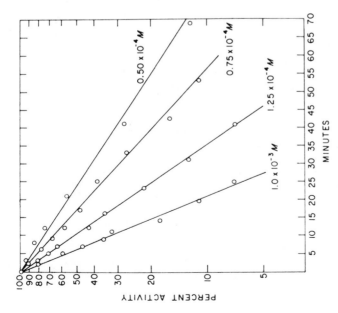

FIGURE 20. The inactivation of bovine trypsin by various concentrations of 1-chloro-3-tosylamido-7-amino-2-heptanone (TLCK). The reactions were performed in 0.05 M Tris, pH 7.0, containing 0.02 M calcium ions at 25°C with 30 μM trypsin. The concentrations of TLCK used are indicated in the figure. (From Shaw, E. and Glover, G., *Arch. Biochem. Biophys.*, 139, 298, 1970. With permission.)

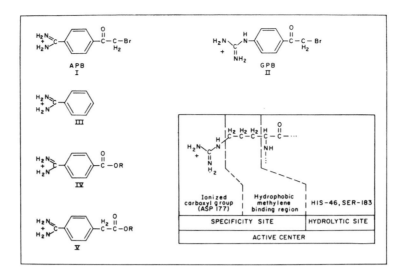

FIGURE 22.   Schematic view of the active center of trypsin indicating the extended arrangement of specificity site carboxyl group, hydrophobic side-chain binding region, and hydrolytic mechanism as deduced from the properties of inhibitors and substrates. Also shown are I(APB), *p*-amidinophenacyl bromide; II(GPB), *p*-guanidinophenacyl bromide; III, benzamidine; IV and V, para-substituted esters derived from benzamidine and phenylguanidine. (From Schroeder, D. D. and Shaw, E., *Arch. Biochem. Biophys.*, 142, 340, 1971. With permission.)

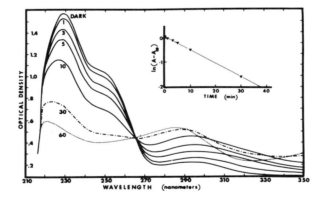

FIGURE 23.   The effect of photolysis on the UV absorption spectra of *m*-azidobenzamidine. The concentration of *m*-azidobenzamidine was 0.187 m*M*. Solid line for 0 (dark), 1, 3, 5, and 10 min of photolysis (performed with a model 457 Micropulser, Xenon Corp., equipped with a Zenon flash lamp, N-725C-WC; energy was delivered at a rate of 14 pulses per sec utilizing a 6 kV potential on the lamp, located 2.5 cm from the sample; samples were contained in Pyrex Petri dishes, 1.5 × 5.0 cm, fitted with a 5.0 cm diameter polystyrene Falcon Petri dish covers to act as filters; this configuration provided efficient thermal equilibration in addition to a very sharp filter cutoff of wavelengths below 295 nm; the entire lamp housing and sample compartment were maintained at 5 to 7°C by enclosure in a small refrigerator). Dashed and dotted lines represent 30 and 60 min, respectively. Photolysis rate (inset) is k_m = 7.4 ± 0.9 × 10^{-5} per pulse. The solvent was 0.1 *M* succinate, pH 6.2. (From DeTraglia, M. C., Brand, J. S., and Tometsko, A. M., *J. Biol. Chem.*, 253, 1846, 1978. With permission.)

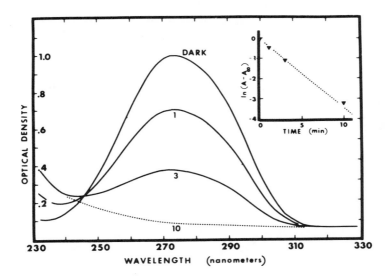

FIGURE 24. The effect of photolysis on the UV absorption spectrum on *p*-azidobenzamidine. The experiments were performed as described under Figure 23 with 0.181 m*M* *p*-azidobenzamidine. Solid lines for 0 (dark), 1 and 3 min of photolysis while the dashed line represents 10 min. Photolysis rate (inset) is $k_p = 4.2 \pm 0.5 \times 10^{-4}$ per pulse. (From DeTraglia, M. C., Brand, J. S., and Tometsko, A. M., *J. Biol. Chem.*, 253, 1846, 1978. With permission.)

Table 2
SELECTED EXAMPLES OF AFFINITY LABELS

Compound	Substrate or Ligand[a]	Enzyme	Residue modified[b]	Ref.
Tosylphenylalanine chloromethyl ketone (TPCK, Tos-PheCH₂Cl) L-1-tosylamido-2-phenylethyl chloromethyl ketone	Tosylphenylalanine methyl ester	Chymotrypsin	His His	1 2
Tosyllysyl chloromethyl ketone (TLCK; TosLysCH₂Cl) 1-chloro-3-tosylamido-7-amino-2-heptanone	Tosyl lysine methyl ester	Trypsin	His	3
	Adenosine 5'-phosphate	Adenylosuccinate lyase		4

Table 2 (continued)
SELECTED EXAMPLES OF AFFINITY LABELS

Compound	Substrate or Ligand[a]	Enzyme	Residue modified[b]	Ref.
BrCH₂ – C(=O) – N(CH₃) – CH(CH₂CH₂C₆H₅) – COOH *N*-Bromoacetyl-*N*-methyl-L-phenylalanine	CH₃ – C(=O) – NH – CH(CH₂C₆H₅) – C(=O) – OR *N*-Acetyl-L-phenylalanine ester	Carboxypeptidase	Glu	5,6
p-Amidinophenacyl bromide	Benzamidine	Trypsin		7
p-Guanidinophenacyl bromide	Benzamidine	Trypsin	Ser	7
m-[*O*-(2-chloro-5-fluorosulfonyl[e] phenylureido) phen-oxybutoxy] benzamidine	Benzamidine	Trypsin Thrombin		8,9 10
DL-α-Bromo-β-(5-imidazoyl)-propionic acid	Imidazole[f]	Horse liver alcohol dehydrogenase, yeast alcohol dehydrogenase	Cys	11
p-Azidobenzamidine	Benzamidine	Trypsin		12
m-Azidobenzamidine	Benzamidine	Trypsin		13
Iodoacetyldiethylstilbestrol		Glutamate	Cys	14

Table 2 (continued)
SELECTED EXAMPLES OF AFFINITY LABELS

Compound	Substrate or Ligand[a]	Enzyme	Residue modified[b]	Ref.
CH₃ / CH–CH₃ / O OH CH₂ O / ClCH₂–C–N–CH–C–OCH₃	Metal affinity	Thermolysin	Glu	15
Br–CH₂–C–NH–CH₂CH₂–O–P–O⊖ ... Bromoacetylcholine	Acetylcholine	Acetylcholine receptor	Cys	16
NHCH₂CH₂NH–C CH₂I ... SO₃H / N-(Iodoacetylamino-ethyl)-5-naphthylamine-1-sulfonic acid		Phosphoenolpyruvate carboxykinase	Cys	17
AMP Dialdehyde (adenosine-5'-monophosphate-2',3'-dialdehyde)	Adenosine-5'-phosphate	Fructose 1,6-bisphosphatase	Lys[d]	18
2-Chloro-3-(5-imidazodyl) propionic acid	Metal affinity	Horse liver alcohol dehydrogenase, yeast alcohol dehydrogenase		19
N-Bromoacetylglucosamine	Glucose	Rat muscle hexokinase II	Cys	20
6-Diazo-5-oxo-D-norleucine	D-Glutamate	γ-Glutamyl transpeptidase		21

Table 2 (continued)
SELECTED EXAMPLES OF AFFINITY LABELS

Compound	Substrate or Ligand[a]	Enzyme	Residue modified[b]	Ref.
N-4-Azido-2-nitrophenyl-γ-aminobutyryl-adenosine diphosphate[j]	Adenine nucleotides	F1-ATPase		22,23
5-[^{125}I]-iodonaphthyl-1-azide	"Hydrophobic" probe	Cytochrome C oxidase		24
5-(4-Azido-2-nitrophenyl)-[^{35}S]-thiophenol	"Hydrophobic" probe	Cytochrome C oxidase		24
N-(2,3-Epoxypropyl)-N-amidinoglycine	Creatine	Creatine kinase	Carbo	25
4-(3-Bromoacetylpyridinio)-butyldiphosphoadenosine	NAD	Horse liver alcohol dehydrogenase	Cys	26
		Yeast alcohol dehydrogenase	Cys	26
		Bacillus stearothermophilus alcohol dehydrogenase	Cys	27

Table 2 (continued)
SELECTED EXAMPLES OF AFFINITY LABELS

Compound	Substrate or Ligand[a]	Enzyme	Residue modified[b]	Ref.
p-Dimethylaminobenzene diazonium fluoride[c]		Acetylcholine esterase		28
5'-*p*-Fluorosulfonyl-benzoyl adenosine	Adenosine-5'-phosphate	Yeast pyruvate kinase	Tyr[f] Lys	29
4-Nitrophenyl-α-D-galactopyranoside[g]	β-Galactoside	Lac carrier protein		30
Azidofluorescein diacetate				31
17β-Hydroxy-4,6-androstadien-3-one[h] (Δ[6] testosterone)	17-β-Hydroxy-5-androstan-3-one	Androgen binding proteins		32
3-Azido-2',7-naphthalene disulfonate	Hydrophilic surface probe			33

Table 2 (continued)
SELECTED EXAMPLES OF AFFINITY LABELS

Compound	Substrate or Ligand[a]	Enzyme	Residue modified[b]	Ref.

$$CH_3-CH_2-C-(CH_2)_{14}-\overset{O}{\overset{\|}{C}}-OCH_2CH_2-NH-\text{(ring)}-N_3$$
(with doxyl oxazolidine ring on the left carbon, and NO_2 on the ring)

2-(2-Nitro-4-azidophenyl)aminoethyl 16-doxylstearate (1,14 Nap) — Ca^{++} ATPase in sarcoplasmic reticulum — 34

Adenosine triphosphate-γ-*p*-azidoanilide — Adenosine triphosphate — Arginine kinase — Cys — 35

$$BrCH_2-\overset{O}{\overset{\|}{C}}-\overset{O}{\overset{\|}{C}}-O^{\ominus}$$

Bromopyruvate — — Flavocytochrome b$_2$ — Cys — 36

Bromoacetyl derivative of pleuromutilin — Tiamulin (pleuromutilin) — Ribosome — — 38

5′-*p*-Fluorosulfonyl-benzoyl adenosine — NADH — 17-β-Estradiol dehyrogenase — —

5′-*p*-Fluorosulfonyl-benzoyl guanosine — Guanosine nucleotide — Phosphoenolpyruvate carboxykinase — — 39

1,2-di-*O*-hexylglycero-3-(ethyldiazomalonamido-ethyl phosphate) — Phosphatidyl ethanolamine — Phospholipase A$_2$ — Val — 40

Table 2 (continued)
SELECTED EXAMPLES OF AFFINITY LABELS

Compound	Substrate or Ligand[a]	Enzyme	Residue modified[b]	Ref.
5-*p*-Fluorosulfonyl-benzoyl-1-*N*⁶-ethenoadenosine	Nucleotide	Pyruvate kinase	Cys	41
N-Bromoacetyl-L-Thyroxine	Thyroxine	Thyroxin-binding globulin	Met	42
(RS)-3-bromo-2-ketoglutarate	α-Ketoglutarate	Isocitrate dehydrogenase		43
5′-*p*-Fluorosulfonyl-benzoyl adenosine (5′-FSO₂BzAdo)	Adenosine triphosphate	Cyclic GMP-dependent protein kinase	—	44
Iodoazidobenzyl-pindolol[j] (±)-1-(indol-4-yloxy)-3-[1-(*p*-azido-*m*-iodophenyl)-2-isobutylamine]-2-propanol	β-Adrenergic agonists	β-Adrenergic receptors	—	45
2-Azido-4-ethylamine-6-iso-propylamine-5-triazine (azido-triazine)	Atrazine	Chloroplast membranes	—	46

Table 2 (continued)
SELECTED EXAMPLES OF AFFINITY LABELS

Compound	Substrate or Ligand[a]	Enzyme	Residue modified[b]	Ref.
[3H]-3'-O-(4-[N(4-azido-2-ni-trophenyl] amino butyryl-adenosine-5'-diphosphate ([3H]-NAP₄-ADP)	Adenosine diphosphate	Adenosine-5-triphosphatase	—	47
[3H]-3'-O-(4-[N-(4-azido-2-ni-trophenyl) amino] butyryl adenosine triphosphate ([3H]NAP₄-ATP)	Adenosine triphosphate	Adenosine-5-triphosphatase	—	47
3-Azido-9[(4-diethylamino)-1-methyl-butylamine]-7-methoxyacridine	Quinacrine[k]	Acridine dye binding sites		48
1-Chloro-2-oxohexanol-6-phosphate	Glucose-6-phosphate	Phosphoglucose isomerase		49
8-Azidoguanosine-5'-triphosphate	Guanosine-5-triphosphate	Guanosine-5-triphosphatase		50

Table 2 (continued)
SELECTED EXAMPLES OF AFFINITY LABELS

a   Model for affinity label.
b   From amino acid analysis unless otherwise indicated.
c   Affinity based on metal chelating ability.
d   Stabilization required (reduction of Schiff base).
e   Photoaffinity labeling by using energy transfer from tryptophanyl residue.
f   Two different sites modified in enzyme.
g   Photoaffinity label. Photolysis with a mercury arc lamp (100 W, type AH-4) with a Corning 0-54 filter to eliminate light below 300 nm).
h   Subject to photoexcitation at 345 nm (n → π* transition) resulting in formation of reactive species.
i   Small extent of modification at Lys, Tyr, and His.
j   Iodohydroxybenzylpindolol, a related compound, was demonstrated to have a k_D with duck erythrocyte membranes of 0.22 n*M*.
k   3-Chloro-9-[(4-(dimethylamino)-1-methylbutyl)amino]-7-methoxyacridine; an antimalarial agent and flavin antagonist.

References for Table 2

1. **Schoellman, G. and Shaw, E.,** Direct evidence for the presence of histidine in the active center of chymotrypsin, *Biochemistry,* 2, 252, 1963.
2. **Morihara, K. and Oka, T.,** Subtilisin BPN': inactivation by chloromethyl ketone derivatives of peptide substrates, *Arch. Biochem. Biophys.,* 138, 526, 1970.
3. **Shaw, E., Mares-Guia, M., and Cohen, W.,** Evidence for an active-center histidine in trypsin through the use of a specific reagent, 1-chloro-3-tosylamido-7-amino-2-heptanone, the chloromethyl ketone derived from Nα-tosyl-L-lysine, *Biochemistry,* 4, 2219, 1965.
4. **Hampton, A. and Harper, P. J.,** The potential of carboxylic-phosphoric mixed anhydrides as specific reagents for enzymic binding sites for alkyl phosphates. Inactivation of an enzyme by an anhydride isosteric with adenosine 5'-phosphate, *Arch. Biochem. Biophys.,* 143, 340, 1971.
5. **Hass, G. M. and Neurath, H.,** Affinity labeling of bovine carboxypeptidase A$_\gamma^{Leu}$ by N-bromoacetyl-N-methyl-L-phenylalanine. I. Kinetics of inactivation, *Biochemistry,* 10, 3535, 1971.
6. **Hass, G. M. and Neurath, H.,** Affinity labeling of bovine carboxypeptidase A$_\gamma^{Leu}$ by N-bromoacetyl-N-methyl-L-phenylalanine. II. Sites of modification, *Biochemistry,* 10, 3541, 1971.
7. **Schroeder, D. D. and Shaw, E.,** Active-site-directed phenacyl halides inhibitory to trypsin, *Arch. Biochem. Biophys.,* 142, 340, 1971.
8. **Bing, D. H., Andrews, J. M., and Cory, M.,** Affinity labeling of thrombin and other serine proteases with an extended reagent, in *Chemistry and Biology of Thrombin,* Lundblad, R. L., Fenton, J. W., and Mann, K. G., Eds., Ann Arbor Science, Ann Arbor, Mich., 1977, 159.
9. **Bing, D. H., Cory, M., and Fenton, J. W., II,** Exo-site affinity labeling of human thrombins. Similar labeling on the A chain and B chain/fragments of clotting α– and non-clotting γ/β–thrombins, *J. Biol. Chem.,* 252, 8027, 1977.
10. **Bing, D. H., Cory, M., and Doll, M.,** The inactivation of human Cl by benzamidine and pyridinium sulfonyl fluorides, *J. Immunol.,* 113, 584, 1974.
11. **Dahl, K. H. and McKinley-McKee, J. S.,** Affinity labelling of alcohol dehydrogenases. Chemical modification of the horse liver and the yeast enzymes with α-bromo-β-(5-imidazolyl)-propionic acid and 1,3-dibromoacetone, *Eur. J. Biochem.,* 81, 223, 1977.
12. **DeTraglia, M. C., Brand, J. S., and Tometsko, A. M.,** Characterization of azidobenzamidines as photoaffinity labels for trypsin, *J. Biol. Chem.,* 253, 1846, 1978.
13. **Tometsko, A. M. and Turula, J.,** Inactivation of trypsin and chymotrypsin with a photosensitive probe, *Int. J. Peptide Protein Res.,* 8, 331, 1976.
14. **Michel, F., Pons, M., Descomps, B., and Crastes de Paulet, A.,** Affinity labelling of the estrogen binding site of glutamate dehydrogenase with iodoacetyldiethylstilbestrol. Selective alkylation of cysteine-89, *Eur. J. Biochem.,* 84, 267, 1978.
15. **Rasnick, D. and Powers, J. C.,** Active-site directed irreversible inhibition of thermolysin, *Biochemistry,* 17, 4363, 1978.
16. **Damle, V. N., McLaughlin, M., and Karlin, A.,** Bromoacetylcholine as an affinity label of the acetylcholine receptor from *Torpedo californica, Biochem. Biophys. Res. Commun.,* 84, 845, 1978.
17. **Silverstein, R., Rawitch, A. B., and Grainger, D. A.,** Affinity labelling of phosphoenolpyruvate carboxykinase with 1,5-I-AEDANS, *Biochem. Biophys. Res. Commun.,* 87, 911, 1979.
18. **Maccioni, R. B., Hubert, E., and Slebe, J. C.,** Selective modification of fructose 1,6-bisphosphatase by periodate-oxidized AMP, *FEBS Lett.,* 102, 29, 1979.

Table 2 (continued)

19. **Dahl, K. H., McKinley-McKee, J. S., Beyerman, H. C., and Noordam, A.,** Metal-directed affinity labelling. Inactivation and inhibition studies of two zinc alcohol dehydrogenases with twelve imidazole derivatives, *FEBS Lett.,* 99, 308, 1979.

20. **Connolly, B. A. and Trayer, I. P.,** Affinity labelling of rat muscle hexokinase type II by a glucose-derived alkylating agent, *Eur. J. Biochem.,* 93, 375, 1979.

21. **Inoue, M., Horiuchi, S., and Marino, Y.,** Affinity labelling of rat kidney γ-glutamyl transpeptidase by 6-diazo-5-oxo-D-norleucine, *Eur. J. Biochem.,* 99, 169, 1979.

22. **Lunardi, J., Lauquin, G. J. M., and Vignais, P. V.,** Interaction of azidonitrophenylaminobutyryl-ADP, a photoaffinity ADP, analog with mitochondrial adenosine triphosphatase. Identification of the labeled subunits, *FEBS Lett.,* 80, 317, 1977.

23. **Lunardi, J. and Vignais, P. V.,** Adenine nucleotide binding sites in chemically modified F1-ATPase. Inhibitory effect of 4-chloro-7-nitrobenzofurazan on photolabeling by arylazido nucleotides, *FEBS Lett.,* 102, 23, 1979.

24. **Cerletti, N. and Schatz, G.,** Cytochrome c oxidase from baker's yeast. Photolabeling of subunits exposed to the lipid bilayer, *J. Biol. Chem.,* 254, 7746, 1979.

25. **Marletta, M. A. and Kenyon, G. L.,** Affinity labelling of creatine kinase by N-(2,3-epoxypropyl)-N-amidinoglycine, *J. Biol. Chem.,* 254, 1879, 1979.

26. **Woenckhaus, C., Jeck, R., and Jörnvall, H.,** Affinity labelling of yeast and liver alcohol dehydrogenases with the NAD analogue, 4-(3-bromoacetylpyridinio)butyldiphosphoadenosine, *Eur. J. Biochem.,* 93, 65, 1979.

27. **Jeck, R., Woenckhaus, C., Harris, J. I., and Runswick, M. J.,** Identification of the amino acid residue modified in *Bacillus stearothermophilus* alcohol dehydrogenase by the NAD^+ analogue 4-(3-bromoacetyl-pyridinio) butyldiphosphoadenosine, *Eur. J. Biochem.,* 93, 57, 1979.

28. **Godner, M. P. and Hirth, C. G.,** Specific photolabeling induced by energy transfer: application to irreversible inhibition of acetylcholinesterase, *Proc. Natl. Acad. Sci. U.S.A.,* 77, 6439, 1980.

29. **Likos, J. J., Hess, B., and Colman, R. F.,** Affinity labelling of the active site of yeast pyruvate kinase by 5′-p-fluorosulfonylbenzoyl adenosine, *J. Biol. Chem.,* 255, 9388, 1980.

30. **Kaczorowski, G. J., LeBlanc, G., and Kaback, H. R.,** Specific labelling of the *lac* carrier protein in membrane vesicles of *Escherichia coli* by a photoaffinity reagent, *Proc. Natl. Acad. Sci. U.S.A.,* 77, 6319, 1980.

31. **Rotman, A. and Heldman, J.,** Azidofluorescein diacetate — a novel intracellular photolabelling reagent, *FEBS Lett.,* 122, 215, 1980.

32. **Taylor, C. A., Jr., Smith, H. E., and Danzo, B. J.,** Photoaffinity labelling of rat androgen binding protein, *Proc. Natl. Acad. Sci. U.S.A.,* 77, 234, 1980.

33. **Moreland, R. B. and Dockter, M. E.,** Preparation and characterization of 3-azido-2,7-naphthalene disulfonate: a photolabile fluorescent precursor useful as a hydrophilic surface probe, *Analyt. Biochem.,* 103, 26, 1980.

34. **Fellmann, P., Andersen, J., Devaux, P. F., le Maire, M., and Bienvenue, A.,** Photoaffinity spin-labelling of the Ca^{2+} ATPase in sarcoplasmic reticulum: evidence for oligomeric structure, *Biochem. Biophys. Res. Commun.,* 95, 289, 1980.

35. **Vandest, P., Labbe, J. –P., and Kassab, R.,** Photoaffinity labelling of arginine kinase and creatine kinase with a ᵧ-P-substituted arylazido analogue of ATP, *Eur. J. Biochem.,* 104, 433, 1980.

36. **Alliel, P. M., Mulet, C., and Lederer, F.,** Bromopyruvate as an affinity label for baker's yeast flavocytochrome b₂. Stoichiometry of incorporation and localization on the peptide chain, *Eur. J. Biochem.,* 105, 343, 1980.

37. **Högenauer, G., Egger, H., Ruf, C., and Stumper, B.,** Affinity labelling of *Escherichia coli* ribosomes with a covalently binding derivative of the antibiotic pleuromutilin, *Biochemistry,* 20, 546, 1981.

38. **Tobias, B. and Strickler, R. C.,** Affinity labelling of human placental 17β-estradiol dehydrogenase and 20α-hydroxysteroid dehydrogenase with 5′-[p-(fluorosulfonyl) benzoyl] adenosine, *Biochemistry,* 20, 5546, 1981.

39. **Jadus, M., Hanson, R. W., and Colman, R. F.,** Inactivation of phospoenolpyruvate carboxykinase by the guanosine nucleotide analogue, 5′-p-fluorosulfonylbenzoyl guanosine, *Biochem. Biophys. Res. Commun.,* 101, 884, 1981.

40. **Huang, K.-S. and Law, J. H.,** Photoaffinity labelling of *Crotalus atrox* phospholipase A₂ by a substrate analogue, *Biochemistry,* 20, 181, 1981.

41. **Likos, J. J. and Colman, R. F.,** Affinity labelling of rabbit muscle pyruvate kinase by a new fluorescent nucleotide alkylating agent 5-[p-(fluorosulfonyl)benzoyl]-1,N^6-ethenoadenosine, *Biochemistry,* 20, 491, 1981.

42. **Erard, F., Cheng, S.-Y., and Robbins, J.,** Affinity labelling of human serum thyroxine-binding globulin with N-bromoacetyl-L-thyroxine: identification of the labelled amino acid residues, *Arch. Biochem. Biophys.,* 206, 15, 1981.

Table 2 (continued)

43. **Hartman, F. C.,** Interaction of isocitrate dehydrogenase with *(RS)*-3-bromo-2-ketoglutarate. A potential affinity label for α-ketoglutarate binding sites, *Biochemistry,* 20, 894, 1981.
44. **Hixson, C. S. and Krebs, E. G.,** Affinity labelling of the ATP binding site of bovine lung cyclic GMP-dependent protein kinase with 5′-*p*-fluorosulfonylbenzoyladenosine, *J. Biol. Chem.,* 256, 1122, 1981.
45. **Rashidbaigi, A. and Ruoho, A. E.,** Iodoazidobenzylpindolol, a photoaffinity probe for the β-adrenergic receptor, *Proc. Natl. Acad. Sci. U.S.A.,* 78, 1609, 1981.
46. **Pfister, K., Steinback, K. E., Gardner, G., and Arntzen, C. J.,** Photoaffinity labelling of an herbicide receptor protein in chloroplast membranes, *Proc. Natl. Acad. Sci. U.S.A.,* 78, 981, 1981.
47. **Lunardi, J., Satre, M., and Vignais, P. V.,** Exploration of adenosine-5′-diphosphate-adenosine 5′-triphosphate binding sites of *Escherichia coli* adenosine-5′-triphosphatase with arylazido adenine nucleotides, *Biochemistry,* 20, 473, 1981.
48. **Mueller, D. M., Hudson, R. A., and Lee, C.-P.,** Azide photoaffinity analogues for acridine dye binding sites, *J. Am. Chem. oc.,* 103, 1860, 1981.
49. **Schnackerz, K. D., Chirgwin, J. M., and Noltmann, E. A.,** Synthesis of 1-chloro-2-oxohexanol 6-phosphate, a covalent active-site reagent for phosphoglucose isomerase, *Biochemistry,* 20, 1756, 1981.
50. **Takemoto, D. J., Haley, B. E., Hansen, J., Pinkett, O., and Takemoto, L. J.,** GTPase from rod outer segments: characterization by photoaffinity labelling and tryptic peptide mapping, *Biochem. Biophys. Res. Commun.,* 102, 341, 1981.

REFERENCES

1. **Jakoby, W. B. and Wilchek, M., Eds.,** *Affinity Labeling, Methods in Enzymology,* Vol. 46, Academic Press, New York, 1977.
2. **Seiler, N., Jung, M. J., and Koch-Weser, J., Eds.,** *Enzyme-Activated Irreversible Inhibitors,* Elsevier/North-Holland, Amsterdam, 1978.
3. **Plapp, B. V.,** Application of affinity labeling for studying structure and function of enzymes, *Meth. Enzymol.,* 87, 469, 1982.
4. **Wofsy, L., Metzger, H., and Singer, S. J.,** Affinity labeling — a general method for labeling the active sites of antibody and enzyme molecules, *Biochemistry,* 1, 1031, 1962.
5. **Baker, B. R.,** *Design of Active-Site-Directed Irreversible Enzyme Inhibitors,* John Wiley & Sons, New York, 1967.
6. **Schoellmann, G. and Shaw, E.,** Direct evidence for the presence of histidine in the active center of chymotrypsin, *Biochemistry,* 2, 252, 1963.
7. **Mares-Guia, M. and Shaw, E.,** Studies on the active center of trypsin. The binding of amidines and guanidines as models of the substrate side chain, *J. Biol. Chem.,* 240, 1579, 1965.
8. **Shaw, E., Mares-Guia, M., and Cohen, W.,** Evidence for an active-center histidine in trypsin through the use of a specific reagent, 1-chloro-3-tosylamido-7-amino-2-heptanone, the chloromethyl ketone derived from Nα-tosyl-L-lysine, *Biochemistry,* 4, 2219, 1965.
9. **Tobias, B. and Strickler, R. C.,** Affinity labeling of human placental 17β-estradiol dehydrogenase and 20α-hydroxysteroid dehydrogenase with 5′-[p-(fluorosulfonyl)benzoyl] adenosine, *Biochemistry,* 20, 5546, 1981.
10. **Kurachi, K., Powers, J. C., and Wilcox, P. E.,** Kinetics of the reaction of chymotrypsin Aα with peptide chloromethyl ketones in relation to its subsite specificity, *Biochemistry,* 12, 771, 1973.
11. **Powers, J. C., Lively, M. O., III, and Tippett, J. T.,** Inhibition of subtilisin BPN′ with peptide chloromethyl ketones, *Biochim. Biophys. Acta,* 480, 246, 1977.
12. **Wold, F.,** Affinity labeling — an overview, *Meth. Enzymol.,* 46, 3, 1977.
13. **Marletta, M. A. and Kenyon, G. L.,** Affinity labeling of creatine kinase by *N*-(2,3-epoxypropyl)-*N*-amidinoglycine, *J. Biol. Chem.,* 254, 1879, 1979.
14. **Meloche, H. P.,** Bromopyruvate inactivation of 2-keto-3-deoxy-6-phosphogluconic aldolase. I. Kinetic evidence for active site specificity, *Biochemistry,* 6, 2273, 1967.
15. **Main, A. R.,** Affinity and phosphorylation constants for the inhibition of esterases by organophosphates, *Science,* 144, 992, 1964.
16. **Brayer, G. D., Delbaere, L. T. J., James, M. N. G., Bauer, C.-A., and Thompson, R. C.,** Crystallographic and kinetic investigations of the covalent complex formed by a specific tetrapeptide aldehyde and the serine protease from *Streptomyces griseus,* *Proc. Natl. Acad. Sci. U.S.A.,* 76, 96, 1979.

17. **Kezdy, F. J., Lorand, L., and Miller, K. D.,** Titration of active centers in thrombin solutions. Standardization of the enzyme, *Biochemistry,* 4, 2302, 1965.

18. **Chase, T., Jr. and Shaw, E.,** Comparison of the esterase activities of trypsin, plasmin, and thrombin on guanidinobenzoate esters. Titration of the enzymes, *Biochemistry,* 8, 2212, 1969.

19. **Livingston, D. C., Brocklehurst, J. R., Cannon, J. F., Leytus, S. P., Wehrly, J. A., Peltz, S. W., Peltz, G. A., and Mangel, W. F.,** Synthesis and characterization of a new fluorogenic active-site titrant of serine proteases, *Biochemistry,* 20, 4298, 1981.

20. **Likos, J. J., Hess, B., and Colman, R. F.,** Affinity labeling of the active site of yeast pyruvate kinase by 5'-p-fluorosulfonylbenzoyl adenosine, *J. Biol. Chem.,* 255, 9388, 1980.

21. **Damle, V. N., McLaughlin, M., and Karlin, A.,** Bromoacetylcholine as an affinity label of the acetylcholine receptor from *Torpedo californica, Biochem. Biophys. Res. Commun.,* 84, 845, 1978.

22. **Schnackerz, K. D., Chirgwin, J. M., and Noltmann, E. A.,** Synthesis of 1-chloro-2-oxohexanol 6-phosphate, a covalent active-site reagent for phosphoglucose isomerase, *Biochemistry,* 20, 1756, 1981.

23. **Tometsko, A. M. and Turula, J.,** Inactivation of trypsin and chymotrypsin with a photosensitive probe, *Int. J. Peptide Protein Res.,* 8, 331, 1976.

24. **Lunardi, J. and Vignais, P. V.,** Adenine nucleotide binding sites in chemically modified F_1-ATPase. Inhibitory effect of 4-chloro-7-nitrobenzofurazan on photolabeling by arylazido nucleotides, *FEBS Lett.,* 102, 23, 1979.

25. **Aiba, H. and Krakow, J. S.,** Photoaffinity labeling of the adenosine cyclic 3',5'-monophosphate receptor protein of *Escherichia coli* with 8-azidoadenosine 3,5'-monophosphate, *Biochemistry,* 19, 1857, 1980.

26. **Goeldner, M. P. and Hirth, C. G.,** Specific photoaffinity labeling induced by energy transfer: application to irreversible inhibition of acetylcholinesterase, *Proc. Natl. Acad. Sci. U.S.A.,* 77, 6439, 1980.

27. **Moreland, R. B. and Dockler, M. E.,** Preparation and characterization of 3-azido-2,7-naphthalene disulfonate: a photolabile fluorescent precursor useful as a hydrophilic surface probe, *Analyt. Biochem.,* 103, 26, 1980.

28. **Kaczorowski, G. J., LeBlanc, G., and Kaback, H. R.,** Specific labeling of the *lac* carrier protein in membrane vesicles of *Escherichia coli* by a photoaffinity reagent, *Proc. Natl. Acad. Sci. U.S.A.,* 77, 6319, 1980.

29. **Taylor, C. A., Jr., Smith, H. E., and Danzo, B. J.,** Photoaffinity labeling of rat androgen binding protein, *Proc. Natl. Acad. Sci. U.S.A.,* 77, 234, 1980.

30. **Abeles, R. H. and Maycock, A. L.,** Suicide enzyme inhibitors, *Acc. Chem. Res.,* 9, 313, 1976.

31. **Walsh, C.,** Recent developments in suicide substrates and other active site-directed inactivating agents of specific target enzymes, *Horizons Biochem. and Biophys.,* 3, 36, 1977.

32. **Light, A.,** The reaction of iodoacetate and bromoacetate with papain, *Biochem. Biophys. Res. Commun.,* 17, 781, 1964.

33. **Shaw, D. C., Stein, W. H., and Moore, S.,** Inactivation of chymotrypsin by cyanate, *J. Biol. Chem.,* 239, PC671, 1964.

34. **Heinrikson, R. L., Stein, W. H., Crestfield, A. M., and Moore, S.,** The reactivities of the histidine residues at the active site of ribonuclease toward halo acids of different structures, *J. Biol. Chem.,* 240, 2921, 1965.

35. **Fruchter, R. G. and Crestfield, A. M.,** The specific alkylation by iodoacetamide of histidine-12 in the active site of ribonuclease, *J. Biol. Chem.,* 242, 5807, 1967.

36. **Lin, M. C., Stein, W. H., and Moore, S.,** Further studies on the alkylation of the histidine residues at the active site of pancreatic ribonuclease, *J. Biol. Chem.,* 243, 6167, 1968.

37. **Holbrook, J. J. and Ingram, V. A.,** Ionic properties of an essential histidine residue in pig heart lactate dehydrogenase, *Biochem. J.,* 131, 729, 1973.

38. **Stevenson, K. J. and Smillie, L. B.,** The inhibition of chymotrypsins A_4 and B with chloromethyl ketone reagents, *Can. J. Biochem.,* 46, 1357, 1968.

39. **Morihara, K. and Oka, T.,** Subtilisin BPN': inactivation by chloromethyl ketone derivatives of peptide substrates, *Arch. Biochem. Biophys.,* 138, 526, 1970.

40. **Segal, D. M., Cohen, G. H., Davies, D. R., Powers, J. C., and Wilcox, P. E.,** The stereochemistry of substrate binding to chymotrypsin A_γ, in *Cold Spring Harbor Symp. Quant. Biol.,* 36, 85, 1971.

41. **Segal, D. M., Powers, J. C., Cohen, G. H., Davies, D. R., and Wilcox, P. E.,** Substrate binding site in bovine chymotrypsin A_γ. A crystallographic study using peptide chloromethyl ketones as site-specific inhibitors, *Biochemistry,* 10, 3728, 1971.

42. **Stein, M. J. and Liener, I. E.,** Inhibition of ficin by the chloromethyl ketone derivatives of *N*-tosyl-L-lysine and *N*-tosyl-L-phenylalanine, *Biochem. Biophys. Res. Commun.,* 26, 376, 1967.

43. **Tsai, I.-H. and Bender, M. L.,** Conformation of the active site of thiolsubtilisin: reaction with specific chloromethyl ketones and arylacryloylimidazoles, *Biochemistry,* 18, 3764, 1979.

44. **Angelides, K. J. and Fink, A. L.,** Cryoenzymology of papain: reaction mechanism with an ester substrate, *Biochemistry,* 17, 2659, 1978.

45. **Connolly, B. A. and Trayer, I. P.,** Affinity labelling of rat-muscle hexokinase type II by a glucose-derived alkylating agent, *Eur. J. Biochem.,* 93, 375, 1979.

46. **Brocklehurst, K. and Malthouse, J. P. G.,** Mechanism of the reaction of papain with substrate-derived diazomethyl ketones. Implications for the difference in site specificity of halomethyl ketones for serine proteinases and cysteine proteinases and for stereoelectronic requirements in the papain catalytic mechanism, *Biochem. J.,* 175, 761, 1978.

47. **Hass, G. M. and Neurath, H.,** Affinity labeling of bovine carboxypeptidase A_γ^{Leu} by *N*-bromoacetyl-*N*-methyl-L-phenylalanine. I. Kinetics of inactivation, *Biochemistry,* 10, 3535, 1971.

48. **Hass, G. M. and Neurath, H.,** Affinity labeling of bovine carboxypeptidase A_γ^{Leu} by *N*-bromoacetyl-*N*-methyl-L-phenylalanine. II. Sites of modification, *Biochemistry,* 10, 3541, 1971.

49. **Takahashi, K., Stein, W. H., and Moore, S.,** The identification of a glutamic acid residue as part of the active site of ribonuclease T_1, *J. Biol. Chem.,* 242, 4682, 1967.

50. **Rasnick, D. and Powers, J. C.,** Active-site directed irreversible inhibition of thermolysin, *Biochemistry,* 17, 4363, 1978.

51. **Schechter, I. and Berger, A.,** On the size of the active site in proteases. I. Papain, *Biochem. Biophys. Res. Commun.,* 27, 157, 1967.

52. **Nishino, N. and Powers, J. C.,** Peptide hydroxamic acids as inhibitors of thermolysin, *Biochemistry,* 17, 2846, 1978.

53. **Bing, D. H., Cory, M., and Fenton, J. W., II,** Exo-site affinity labeling of human thrombins. Similar labeling on the A chain and B chain/fragments of clotting α- and non-clotting γ/β thrombins, *J. Biol. Chem.,* 252, 8027, 1977.

54. **Cory, M., Andrews, J. M., and Bing, D. H.,** Design of exo affinity labeling reagents, *Meth. Enzymol.,* 46, 115, 1977.

55. **Bergmann, M., Fruton, J. S., and Pollok, H.,** The specificity of trypsin, *J. Biol. Chem.,* 127, 643, 1939.

56. **Inagami, T. and Hatano, H.,** Effect of alkylguanidines on the inactivation of trypsin by alkylation and phosphorylation, *J. Biol. Chem.,* 244, 1176, 1969.

57. **Petra, P. H., Cohen, W., and Shaw, E. N.,** Isolation and characterization of the alkylated histidine from TLCK inhibited trypsin, *Biochem. Biophys. Res. Commun.,* 21, 612, 1965.

58. **Shaw, E. and Glover, G.,** Further observations on substrate-derived chloromethyl ketones that inactivate trypsin, *Arch. Biochem. Biophys.,* 139, 298, 1970.

59. **Schroeder, D. D. and Shaw, E.,** Active-site-directed phenacyl halides inhibitory to trypsin, *Arch. Biochem. Biophys.,* 142, 340, 1971.

60. **DeTraglia, M. C., Brand, J. S., and Tometsko, A. M.,** Characterization of azidobenzamidines as photoaffinity labels for trypsin, *J. Biol. Chem.,* 253, 1846, 1978.

61. **Staros, J. V., Bayley, H., Standring, D. N., and Knowles, J. R.,** Reduction of aryl azides by thiols: implications for the use of photoaffinity reagents, *Biochem. Biophys. Res. Commun.,* 80, 568, 1978.

INDEX